The
NEURO-LINGUISTIC
PROGRAMMING
Workbook

Judy Bartkowiak

Teach®
Yourself

The NEURO-LINGUISTIC PROGRAMMING Workbook

Judy Bartkowiak

First published in Great Britain in 2012 by Hodder & Stoughton. An Hachette UK company.

This edition published in 2017 by John Murray Learning

British Library Cataloguing in Publication Data: a catalogue record for this title is available from the British Library.

Library of Congress Catalog Card Number: on file.

ISBN: 978 1 47365 962 9

eISBN: 978 1 47365 910 0

1

The publisher has used its best endeavours to ensure that any website addresses referred to in this book are correct and active at the time of going to press. However, the publisher and the author have no responsibility for the websites and can make no guarantee that a site will remain live or that the content will remain relevant, decent or appropriate.

The publisher has made every effort to mark as such all words which it believes to be trademarks. The publisher should also like to make it clear that the presence of a word in the book, whether marked or unmarked, in no way affects its legal status as a trademark.

Every reasonable effort has been made by the publisher to trace the copyright holders of material in this book. Any errors or omissions should be notified in writing to the publisher, who will endeavour to rectify the situation for any reprints and future editions.

This book is for information or educational purposes only and is not intended to act as a substitute for medical advice or treatment. Any person with a condition requiring medical attention should consult a qualified medical practitioner or suitable therapist.

Typeset by Cenveo® Publisher Services.

Printed and bound in Great Britain by CPI Group (UK) Ltd., Croydon, CR0 4YY.

John Murray Learning policy is to use papers that are natural, renewable and recyclable products and made from wood grown in sustainable forests. The logging and manufacturing processes are expected to conform to the environmental regulations of the country of origin.

Carmelite House

50 Victoria Embankment

London EC4Y 0DZ

www.hodder.co.uk

Contents

Meet the author

Judy Bartkowiak runs a therapy practice offering one-to-one NLP therapy and workshops from her home in Burnham in Buckinghamshire. Judy is a qualified NLP Business Practitioner, Sports Practitioner, Kids Practitioner, Master Practitioner and NLP Trainer. She trained with Sue Knight in 2001, then with Gemma Bailey and Jeremy Lazarus, returning to Sue Knight for her trainer's training in 2011. Judy is the author of *Be a happier parent with NLP* and the *Engaging NLP* series of workbooks, which are for parents, children, tweens, teens, new mums, returners to work, teachers and people at work.

While her four children were young, Judy ran a Montessori School from home and then returned to working freelance as a children's market researcher before deciding to focus on NLP and writing. She writes children's fiction under the pen name JudyBee. She is also working on a novel and extending the *Engaging NLP* series. Judy is a keen sportswoman and enjoys tennis, hockey, skiing and swimming.

Introduction

During the 1970s Richard Bandler and John Grinder began research into how effective people communicate. They studied the work of leading therapists of their day, including Virginia Satir (family therapist), Fritz Perls (Gestalt therapist) and Milton Erickson (clinical hypnotherapist), and linguists Alfred Korzybski and Noam Chomsky. Their intention was to code exemplars of excellence in communication so that this code could be passed on and replicated by anyone. They called this coding process neuro-linguistic programming (NLP). It codes how we think (neuro), how we communicate (linguistic) and how we get the results we get (programming).

What they developed then has evolved over the years and become a way of thinking based on choices and the notion that there is a positive intention underpinning all action that can be applied in all walks of life. Whatever you do for a living, whatever life stage you are at and wherever you live, you will find that NLP will offer you a fresh and different way of connecting with the people you care about and enable you to discover more about yourself, how you tick and how what you do well in one sphere of your life can be transferred and applied in other areas. Not only can you learn about yourself but you can also learn how to acquire the skills of others you admire and add them to your own skill set.

NLP is experiential, in that you will learn it by experiencing it, and that is what we are inviting you to do in this workbook.

For a more 'meaty' definition of NLP, see Chapter 1.

→ How to use this workbook

You have bought this book so we are assuming that you are curious about NLP. Maybe you've heard about it at work or on TV, or maybe you've already read something on the subject and want to know more about how you can use it in your own life. You may have realized that if you always do what you've always done, you will always get what you've always got, and so you feel that now might be the time to make some changes so you can get something different.

Be prepared to get involved in this workbook!

This is *not* a textbook with lots of jargon-filled pages of theory about how 'other people' have used NLP in the workplace to further their sales or management career, implement change management programmes or become inspiring leaders. There are plenty of those on the bookshelves, dusty and unused. Instead, we want you to use this book and apply the exercises to you personally in every aspect of your life. Be curious and be prepared to 'have a go', get involved and keep an open mind about what you can learn and apply in your own life. Let go of the 'should', 'must' and 'ought to' and embrace the desire for self-discovery and the option for change.

This is *your* workbook, so make it yours by keeping it somewhere handy so you can dip into it whenever you need to. You will need a pen, so choose a special one that you enjoy writing with and keep it with your workbook. It contains questions and quizzes and places for you to note what you've learned from each section. This will help you move on to the next one, taking the learning from the last section as you journey through the book. Feel free to make your own personal notes and reminders in the book, and underline the points you think are most relevant to you. Sometimes you may read something that reminds you of a friend or relation, so write their name by the side of that part.

Giving up your banana

Let me tell you a story.

There are still places in the world where they hunt monkeys. They lay traps for them. They sink a cage into the ground and put a banana in it. The bars are just wide enough for the monkey to put its hand in to reach the banana but when he has hold of the banana he cannot get his hand back out through the bars. He has to drop the banana to be free.

The trapper comes along and even though the monkey knows that if he doesn't let go of the banana he will be caught, he cannot let go of it. He has to keep hold of the banana. He cannot see that the banana is not helping him achieve his bigger goal of staying alive; he can only think about what he wants right now.

That banana is your beliefs and behaviour.

Are you ready to give up your 'banana' and change?

→ # The core principles of NLP

Before you start on this journey of self-discovery you'll need to pack a metaphorical bag and take with you some beliefs that will help you along the way. Each one is a belief that you need for the journey. Beliefs aren't facts, they are views that we can choose to hold and use as if they are facts. I invite you as we prepare for the journey to read each of these beliefs set out in this chapter and be prepared to hold each one as a fact because they will be a very useful resource for you. In NLP textbooks you will find these are referred to as the NLP **presuppositions**. This means that we are supposing them to be true before we start.

The NLP presuppositions form the framework that underpins the tools and techniques you will learn in this book. They were compiled by John Grinder and Richard Bandler when they developed neuro-linguistic programming, drawing on the therapies they studied in their pursuit of the study of the structure of excellence.

You may find that some of them challenge beliefs you currently hold. This is to be expected because they are new to you, so read them and ask yourself how holding that particular belief would be helpful to you and allow yourself to act as if it were true for the purpose of moving forward in your life with a different choice. What would it enable you to do differently if you believed it to be true?

1 We already have all the resources we need

Somewhere in your life you are already using the skills and qualities you need for whatever is challenging you now.

Focus on what you do well. What do you do 'with excellence' and when you do that thing, what does that also mean you can do well? Write a list for yourself as with the example shown below.

What I do well	What's the skill?	What that also means I can do well
Make good scrambled eggs	Patience	Wait in line, learn new things, teach

What I do well	What's the skill?	What that also means I can do well

Are you being modest? I expect you are! Now think about what other people say you do well and what that means. Again, I've given you an example.

What others say I do well	What's the skill?	What that also means I can do well
Speak French	Listen, remember and repeat	Give a talk, learn lines in a play

It may surprise you to learn that even things we think of as faults or negative qualities can actually be skills! What do you do badly but with excellence? Here are some examples.

What I do badly – well	What's the skill?	What does that also mean I can do well?
Be stubborn	Persistence	Compete in sport

What I do badly – well	What's the skill?	What does that also mean I can do well?

What others say I do badly	What's the skill?	What that also means I do well
Speak about my skills	Put others first	Coaching

By now you should have a long list of skills down the centre of the tables, so I hope you are now convinced that you have all the resources you need for this NLP journey.

2 There is no failure, only feedback

We learn from our mistakes and move on. Each so-called 'failure' has a positive intention for us to learn from it and do something different next time. Behind every behaviour or experience is a positive intention for us to learn from it. Imagine if we gave up trying to walk as toddlers because we kept falling down: we'd never learn to walk!

How do you respond to feedback? It's our choice how we manage feedback. Tick the responses you usually make and sometimes make.

	Usually	Sometimes
Shrug my shoulders and forget about it	❏	❏
Ignore it – it didn't happen	❏	❏
It was someone else	❏	❏
What – me?	❏	❏
Tell myself I'm useless	❏	❏
Give up	❏	❏
Analyse what went wrong	❏	❏
Learn from it	❏	❏

During that exercise, you will have brought to mind things you've done that you feel bad about. Maybe the choice you made about how to deal with the feedback wasn't as resourceful as it could have been. Maybe you've even made the same mistake several times. Let's be quite specific now and write down something you can remember well, something that went badly wrong. Write about it below.

_____

_____

_____

_____

_____

Now, what was the learning? Let's assume this bad experience has a positive intention for you. It wants to teach you something so you make a different choice the next time you are in that situation. Maybe the situation was so bad you have avoided repeating it, in which case learning how to make a different choice will open up possibilities for you that you have closed off until now. Write down below what you feel is the positive outcome in terms of the learning.

_____

_____

_____

Write what you will do differently next time.

_____

_____

_____

Let me introduce you to the **feedback sandwich.**
You can use the feedback sandwich in any situation
as a way to extract the learning and put it in a
positive frame, focusing on the positive intention
behind the experience.

1 Start with what went well. What are you pleased about, what was good?

2 Now write down what could have been better. What do you need to do
   more of and what do you need to do less of? We are looking here for
   change and what could be improved.

3 The last part of the sandwich is the overall positive output. Overall, what
   went well?

Use the process throughout the workbook to give yourself feedback as you
reach the end of each chapter. There will be a specific opportunity to do
that, but it is helpful to bear it in mind after each stage of the learning.

3 If one person can do it, anyone can do it

We can learn from how others do something, copy the structure and achieve
it for ourselves. This is called modelling. Modelling in this context is not
about fashion and showing off clothes but is in the craft context of modelling
clay. When potters throw a pot, they have in mind how they want the pot to
look. We cannot tell, simply by watching them work, what the pot will be like
when they have finished. We need to do more than watch them at work.

We would need to ask them what they have in their head, their thoughts about what they are making. It is the same with NLP modelling. By asking questions that are 'clean' and free of our own assumptions, we ask what beliefs our model has about what he is doing. If we want to replicate that pot, that behaviour, we need to replicate the thought process as well as what we see and hear.

You will have plenty of opportunity to do this in a later chapter. We will also be practising 'clean language'.

4 The meaning of the communication is the effect it has

It's not what you say or do that matters but the effect it has on those who receive it. How often have we been misunderstood because what we meant to say or do hasn't quite come across as we intended? Even if we retrack and try again, the initial response cannot be forgotten. We can ensure that we communicate what we intend to be received by understanding more about the way people process communication. This will be covered in detail in Chapter 2 but, for now, just note down below an occasion when you inadvertently communicated something that was not received well and how you could have avoided this happening.

What happened	What I could have done differently

This sort of miscommunication sometimes occurs when we jump to conclusions, respond too quickly without thinking or make an assumption about what someone meant, based on our own beliefs rather than theirs. It can also happen when we are tired or feeling grumpy about something unrelated to our current situation. For example, we might view a queue at the supermarket checkout as an opportunity for a chat with a fellow shopper when we are not in a hurry but as an irritating situation when we are. The queue is the same; the difference is the meaning we place on it. Only *we* know that meaning.

5 Everyone has a unique view of the world

How you see the world is unique to you. It will be different from everyone else's, so be curious about other people's 'map' of the world. We all have a different upbringing and our view of the world will be based on this, as well as our age, gender, life stage, work situation, education and so many other things. Not surprisingly then, our interpretation of any given situation will be different. Think of a situation in your life when you realize that someone else had a different map from yours. Write it down here.

_____

_____

_____

_____

6 The person with the most flexibility in thinking and behaviour has the greatest influence

You have choices in what you say and do. You can make different choices to get the results you want. The more choices you have, the more flexibility you have and the more chance you have of getting what you want.

How often do you use the word 'should'? This is a 'toxic' word, in that it is unresourceful. It is the enemy of choice. Here are just a few examples.

What are your 'shoulds'? Write them in the thought bubbles below.

If we change them to 'coulds', how would they look?

Let's rewrite them. Use the bubbles below to restate these as choices by replacing 'should' with 'could'.

Write down in the box below how your feelings have changed once you express these as choices.

7 If you spot it, you've got it

What you recognize in others, you have yourself. That's how you can recognize it. Do you find yourself wishing you were like other people? Do you envy their abilities in some way? Do you want their qualities?

Well, just STOP!

Ask yourself in what way you have that quality, that ability, that skill.

Write down here what you really like about your best friend, a close colleague or your partner.

_____

_____

_____

_____

And now write down in the box below in what way you also have those qualities.

But equally, if someone is annoying you or irritates you by their behaviour, ask yourself, in what way do I do that too? It may not be a precise copy of the behaviour but the structure of it. For example, say someone is talking very loudly on their mobile next to you in the train. You may not do that exact thing but do you in some way also disregard other people's needs? How?

_____

_____

_____

_____

8 Mind and body are ONE

You can affect what you do by thinking differently. Once you believe you can do something, you will be able to do it. Similarly, if you decide you can't do something, you won't be able to do it. You are the one making the choice to be able to do something or not. It's all happening in your head before you move your body.

If you play a sport, you will be very aware that your state of mind can affect whether you play well or badly. If you do a few poor shots and feel annoyed with yourself, you may well go on to play more bad shots, whereas freeing your mind of judgements and trusting your body will bring you better results.

Here's a quick and easy experiment for you to do.

You will need a full-length mirror for this one.

With your back to the mirror, get yourself into a state where you imagine you are amazingly handsome/beautiful. Think about a time when you looked fantastic, fit and healthy, and you were wearing clothes that look superb on you. Your hair looks really good and your face looks radiant and healthy. You look half your age and everyone says how great you look. Adjust your body posture so you really get that super feeling throughout your body.

Now turn around and look at yourself in the mirror.

Don't you look gorgeous?

Notice the position of your head and shoulders, how your arms are placed and your legs. Give yourself marks out of 10 for how wonderful you look. Write it in the circle.

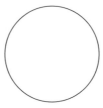

Now turn your back on the mirror again and think about feeling old, overweight, wearing clothes that don't do you justice, and with hair that needs washing and a good cut and wearing no makeup; you feel really quite low.

Now turn around and look at yourself in the mirror.

Don't you look dreadful?

Notice the position of your head and shoulders again. What are your arms and legs doing? Give yourself a score out of 10 for how you look. Write it in the circle.

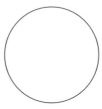

What are your scores? They are different, aren't they? And yet your body is the same body. The only thing that has changed is your mind.

Write down in the space below how you could use this discovery about the mind–body connection in your own life.

_____

_____

_____

_____

9 There is a solution to every problem

Be creative, make different choices and challenge limiting beliefs (beliefs that hold you back from achieving what you want) and you will find the solution.

> *If you always do what you've always done then you'll always get what you've always got.*

Write down a problem you have in the box below.

Now look at it and underline all the negatives and limiting beliefs, such as:

- couldn't
- had to
- shouldn't
- mustn't
- impossible
- don't
- tried

How many did you use? ☐

Now write out the problem again, this time emphasizing what you want to do rather than what you don't want or can't do. Once you focus on what you do want or can do, the solution is much clearer, isn't it?

10 The way to understand is to do

It is by 'doing' that we learn, rather than just by studying the theory. So use this workbook as a way to experience NLP at work in your life. Complete the exercises in each chapter and get involved. That way you will see changes.

Exercise 1

This chapter has looked at the NLP presuppositions.

→ Which of these did you find easiest to take on board?

_____

_____

→ Why is that?

_____

_____

→ Which was the hardest?

_____

_____

→ What limiting belief do you need to change in order to take on this presupposition?

_____

_____

→ Where has that belief come from?

_____

_____

→ Is it still relevant today for you?

_____

_____

→ Are you prepared to do that?

_____

_____

1 The benefits of NLP

This chapter tells you more about how NLP can make a difference in your life. One of the principles of NLP is 'If you always do what you've always done you'll always get what you've always got' so in this first chapter we explore choice. We look at how we process experiences and interactions and how we choose to respond. To demonstrate how NLP can be employed across all aspects of your life, you will do exercises in choice to introduce you to different ways of being you at work, in sport and socially.

→ An overview of NLP

As we have seen, NLP stands for **neuro-linguistic programming**.

Neuro is what goes on in our head. There are two sides of the brain; the left side handles all things logical and analytical and is the area that provides our reasoning and language. The right side of the brain is the creative side, which allows us to be intuitive and recognize patterns such as faces and tunes.

Another aspect of the brain is the conscious and unconscious mind. The conscious mind is logic driven like the left side of the brain, and it is with our conscious mind that we make decisions and choices. The unconscious mind is, meanwhile, controlling our physiology, all our body functions, allowing us to breathe and move. The conscious mind can control the unconscious; for example, it can tell it that you are scared of spiders or giving a talk and the unconscious mind will produce a fear response because it has no ability to reason.

Would it surprise you to know that most of our thought processes are at the unconscious level? Our conscious mind can only cope with between five and nine pieces of information at any one time, whereas our unconscious mind processes millions of bits of data every second. Therefore the unconscious mind is what determines behaviour. NLP uses these external behavioural clues to see beyond the activities of the conscious mind in order to understand the processes running at the unconscious level.

The reason NLP is so successful in so many different areas of life – at work, at home, in sport and for our health – is that NLP techniques bypass the conscious mind and work directly at the unconscious level which controls behaviour.

The core principle of NLP – 'If you always do what you've always done you will always get what you've always got' – means is that if your current behaviour is unconsciously not getting you what you want in life then, by making this unconscious behaviour conscious, you can make different and more resourceful choices and get a more satisfactory result.

Here is a quick and easy exercise that demonstrates how, by making a different choice, you can influence how you feel.

Exercise 2

Look out of the window – what do you see, hear and feel?

➜ What do you see?

_____

_____

➜ What do you hear?

_____

_____

➜ What do you feel?

_____

_____

Write down three things that make you feel good about what you have observed, heard and felt.

1 _____

_____

2 _____

_____

3 _____

_____

Now write down three things that make you feel bad about what you have observed, heard and felt.

1 _____

_____

2 _____

_____

3 _____

_____

You see, you have a choice. What you have out of your window can be thought about in at least two different ways and yet the view is exactly the same. The only difference between your two lists is how you have decided to think about them.

You can do this exercise with any situation. Here are a few situations where you could alternately focus on what is going well and what is not going well, to notice that the event is the same. The only difference is the meaning you choose to give it.

* **When you are running or exercising**
* **During a meeting at work**
* **Getting the children to bed**

The word we use to describe this process is called 'reframing' and, like the window frame in our exercise it is about changing the frame, changing the way you look at something.

Neuro is how we think, our beliefs and values, what we bring to our adult world from what we learned and experienced as children and how we have modified these as we move through adulthood experiences. Our thinking process is not set in stone. You will learn in this workbook how to change the way you think and understand how others think so that you can change your choices and get a different outcome. Some people are reluctant to make changes and believe that they are who they are; 'like it or lump it', so to speak. That's all very well if you are getting the results you want, but if you aren't, and feel you could be happier, more successful, more popular, and more effective at work, home or socially, you will enjoy the exercises in this book and soon experience changes that will give you choices in how you think or perceive.

Can you think of people you know who always seem to see the positive in situations? They are people we describe as having their 'glass half full'. Whatever befalls them, they can see a positive angle. They have a 'positive frame'. Others have a 'glass half empty' and see the negative in situations. They have a 'negative frame'.

Another way to change how you think about a situation is to 'disassociate'. To do this, you need to become 'a fly on the wall' and consider the situation as someone else might experience it. Here is the previous exercise again, but this time act as if you are a child. It may help to decide on a specific child and visualize yourself as that child at that age, with that particular personality.

Exercise 3

You are that child. Now look out of the window again. What do you think of what you see, hear and feel now?

→ What do you see?

_____

_____

➜ What do you hear?

_____

_____

➜ What do you feel?

_____

_____

Write down three things that make you feel good about what you have observed, heard and felt.

1 _____

_____

2 _____

_____

3 _____

_____

Now write down three things that make you feel bad about what you have observed, heard and felt.

1 _____

_____

2 _____

_____

3 _____

_____

What do you notice about the two experiences from these exercises?

_____

_____

_____

_____

Disassociating gives you the chance to distance yourself from the event and be curious about how someone else might perceive it. Remember from the introduction that 'the map is not the territory': there is more than one way to view any event and you can choose the way that is most resourceful for you.

As you work through this book you will become more aware of how what you think influences the results you get, so learning about how you think is the key to successful results.

Linguistic is how we speak and write: the words we use and how we say them, our tone of voice, volume, pace and pitch. Today the speed of communication is fast. We can email or text someone on the other side of the world and get a response in seconds. How important is it, then, to ensure that what we say is a true reflection of what we mean and that it connects in a positive way with the recipient of the message? We have all had emails and texts that have annoyed us because of the words used and we have made assumptions about the sender based on those words. Time does not always allow us to explain what we meant or undo the misunderstanding, and yet global deals and business can be adversely affected by a poor choice of words.

We can consciously change the way we speak in order to get a different response. Remember that the meaning of the communication is the effect it has, so by changing how we use language we can have a different effect. In Chapter 2 we will find out whether your preferred mode of processing information is visual, auditory or kinaesthetic; these reflect language preferences that will determine your choice of words. In Chapter 3 we take this further with a study of the NLP **meta-programmes** that determine how you express yourself, and how – by matching preferences with the person you are communicating with – you can achieve enhanced rapport and influence.

The first thing we notice about someone is their non-verbal communication before they have even uttered a word. We make assumptions about them based on:

- how they stand or sit
- the level of eye contact they give
- how they look
- their mannerisms
- what they are wearing
- their facial expression

Exercise 4

Be aware of what you are communicating non-verbally right now. If I were sitting opposite you, what would I see and what assumptions might I make about you?

How could you give me the impression you'd like me to have of you by altering your appearance and body language?

_____

_____

_____

Linguistic also refers to how we speak to ourselves, our inner voice. How you speak to yourself can be unravelled using NLP as, just like your external voice, it has a structure which we can understand. We'll have a go at that now.

Exercise 5

How does your inner voice sound to you? What does it say?

Write down an example of what your inner voice says to you in the box below.

→ Is it a parent?

→ Is it someone from the past?

→ Is it someone else talking to you and using your name?

→ Does it sound critical?

→ Is it negative?

If you've answered 'yes' to any of these, rewrite the sentence from above in the first person; for example 'I am.........' By doing this you are taking ownership of the criticism and joining in yourself with the inner critic.

_____

_____

_____

It probably sounds quite stark and absolute now, doesn't it? So what we need to do is open up the possibilities by adding the word 'and' after the sentence. This gives you the opportunity to respond as yourself to the criticism and give yourself choices about what to do next. It also calms down the inner voice and takes some of the absolutism out of it. Rewrite the sentence above now in the space below and complete it by adding 'and' and say how you will respond, what you will do about that criticism. Avoid being defensive but proactively take charge of it and make changes.

_____

_____

_____

and _____

Use this process every time you get an annoying inner voice.
_____

Programming is the way we put these patterns of thinking, language and behaviour together to get the results we do, whether they are good results or bad ones. Just as a computer runs programmes to perform functions, so do we. These programmes are strategies and by understanding them we give ourselves choices in everything we do.

Let's have a look at some of your strategies now.

Exercise 6

Imagine you have to make a decision about whether to apply for a new job. What are the thought processes, internal dialogue, actions and conversations that you have along the way to making the decision? Write them out as 10 steps below.

1 _____

_____

2 _____

_____

3 _____

_____

4 _____

_____

5 _____

_____

6 _____

_____

7 _____

_____

8 _____

_____

9 _____

_____

10 _____

_____

Now imagine you are starting a new business. What are the steps you would take?

1 _____

_____

2 _____

_____

3 _____

_____

4 _____

_____

5 _____

_____

6 _____

_____

7 _____

_____

8 _____

_____

9 _____

_____

10 _____

_____

Looking at these two strategies, what do you notice? What are the similarities?

_____

_____

_____

_____

The important thing to notice is whether the strategy works as well as it could or whether by changing one part of it you may have a different and more successful result. What step might you want to change to get a different result?

_____

_____

_____

_____

NLP helps us to make sense of our experience (both conscious and unconscious) and put into practice new patterns of behaviour and language to achieve what we want in all areas of our lives.

Remember that in the introduction you learned that whatever you need, you already have in some way. By learning about NLP and applying the techniques, you can access the resources you need and take them into the part of your life where you need it now.

Focus points

- How you think, what you say and what you do gives you the results you get in whatever you are doing. These happen unconsciously as you use patterns or habits that have developed since childhood and are underpinned by your beliefs.
- We are starting to learn how we can change our beliefs, reframe external cues and make conscious choices about how to respond to them.

- A way to introduce ourselves to other options is to disassociate and view the situation from another viewpoint, using another map of the world.
- Being curious about how we do what we do and how others do what they do is an important part of NLP and opens up possibilities for change to get different, better results.
- You have started to understand that in order to make changes we need to understand the strategies we run now.

 # Where to next?

In the next chapter, you will start to learn how you process your thoughts and feelings. We are starting on a journey of self-discovery that will enable you to understand why you get on with some people better than others, why you like doing the things you do and what you can do to ensure that you communicate in a way that best engages with the person you are talking to. Engaging with people is called 'being in rapport' and Grinder and Bandler found, in their research into the structure of excellence, that rapport was fundamental in all aspects of life and the basis upon which excellence is built. To be in rapport we need to understand how we process information ourselves so that we can recognize the patterns in those we interact with. Once we've done this, we match them to achieve rapport. We match them, too, in our non-verbal communication and body language, which we will expand on in the next chapter.

How being visual, auditory or kinaesthetic affects our choices

In the previous chapter you learned how you can make your unconscious conscious and thereby increase your awareness of how you think, how you look and what you say to yourself, in order to make different choices that will give you different results. This chapter delves specifically into the way you prefer to represent what you see, what you hear or what you do and feel. In NLP it is called our 'internal representation'. A quiz and some exercises will help you discover which method you use, and then you'll learn what this means for you and how you can use your representation for motivating self-talk and influencing others.

➡ VAK and communication

Have you noticed how, with some people, you instantly seem to get along? You seem to talk the same language, almost finishing each other's sentences? With others you struggle to figure out what they mean and the conversation is a bit stilted and awkward. The reason for this is that we communicate in three different ways – visually, auditorily and kinaesthetically (often known as VAK). Each of us has a preferred style, which will be one of these. When we talk to someone using the same representation, the conversation flows easily but can be more awkward when we talk to someone with a different one. Those who communicate most effectively can match the representation of the person they are communicating with. Do this quiz and find out how you communicate best.

Exercise 7

1 When someone is talking to you, what do you do?

   a Look at them ☐

   b Listen to what they're saying ☐

   c Feel you are getting to know them ☐

2 When you have a party, how do you send the invitation?

   a By post or by hand, as a beautifully designed invitation ☐

   b By phoning ☐

   c Via text or email ☐

3 What will be the most important aspect of the party?

   a What you will wear and the décor ☐

   b The music ☐

   c The atmosphere ☐

4 How do you usually catch up on the news?

   a Read the newspaper ☐

   b Listen to the radio ☐

   c Use the Internet ☐

5 When you spell the word 'accommodation', what do you do first?

   a Imagine it written down ☐

   b Sound it out ☐

   c Write it down ☐

6 **When you are in the car driving somewhere, what do you pay attention to?**

   a What I see out of the window ❏

   b What I can hear on the radio ❏

   c How I am driving ❏

7 **When you look up Internet sites, what are you most drawn to?**

   a An attractive home page ❏

   b Video ❏

   c Good functionality ❏

8 **When driving somewhere unfamiliar, what do you find most helpful?**

   a A map or visual display on your sat nav ❏

   b Listening to instructions on the sat nav or asking for directions ❏

   c Trusting your instincts and using the signs ❏

9 **When you go to a movie, what do you notice most?**

   a The pictures on the screen ❏

   b The soundtrack ❏

   c The experience of going to the movie ❏

10 **When you think back to your schooldays, what comes to mind?**

   a Images of you at school with your friends ❏

   b What was said to you, the sounds of you and your friends ❏

   c How you felt about your schooldays ❏

- **Number of 'a' answers** ❏
- **Number of 'b' answers** ❏
- **Number of 'c' answers** ❏

You will have noticed a pattern in answering these questions. You will have mostly answered a, b or c. Although we use all our senses, we tend to prefer one of them.

Visual people

If you mostly answered 'a' you will prefer to focus on what you see. You are 'visual'. You probably pay attention to how you look, what colours you put together and you like to look your best because it is important to you. How others look will also be important and you will notice what others are wearing, whether they've changed their hairstyle, had their hair cut or have a new outfit. You notice everything about your environment as well, so you like to arrange it where you can, to look attractive. Maybe you personalize your desk area at work and pay attention to how your home looks, the décor and the way you arrange the furniture. When you are in other people's houses, you mentally rearrange things and notice the colours and textures they have used. If your home is untidy and cluttered you will feel uncomfortable. At work, the visual appearance of the office will affect how happy you are working there.

Visual people are usually quite artistic and enjoy creative hobbies such as painting and drawing, dressmaking, cooking, writing, pottery and other decorative arts. They often take jobs that have a visual aspect to them such as designing or creating things.

Maybe you point things out to people round you: 'Look at that!', 'Did you see what she was wearing?', 'What an amazing sunset', 'What pretty flowers.' What you see will feature in whatever you do and how you think, so if I say a word to you like 'coffee' you will picture the coffee, whether it be in the mug, the coffee machine or you sitting drinking it.

When learning something new, a new language or revising for a test, learning a new process at work, visual people like to see instructions written down or write them down themselves to refer to.

Visual people tend to look up with their head and eyes as they talk because they are recalling or constructing images. They also talk fast as they try to convey the images verbally and use a reasonably high pitch.

Here are some of the words and phrases to listen out for that suggest a visual representation.

Do you see what I mean?   Let's focus on the key points.   Perhaps you could enlighten me?   In hindsight...

Exercise 8

Now it's your turn to think of some. Write them in the speech bubbles.

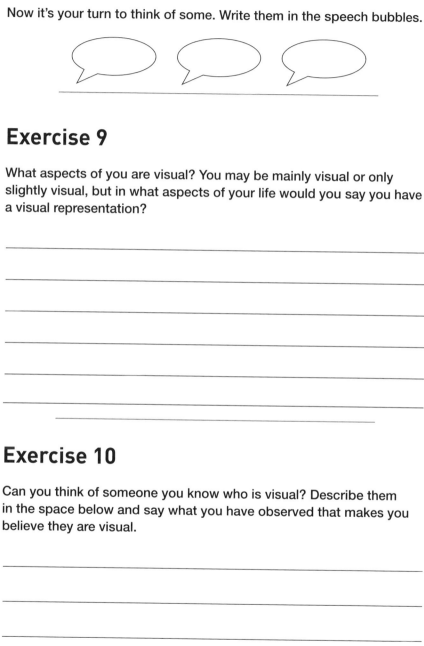

Exercise 9

What aspects of you are visual? You may be mainly visual or only slightly visual, but in what aspects of your life would you say you have a visual representation?

_____

_____

_____

_____

_____

_____

Exercise 10

Can you think of someone you know who is visual? Describe them in the space below and say what you have observed that makes you believe they are visual.

_____

_____

_____

_____

Auditory people

If you mostly marked 'b' in the quiz, this means that your representation is 'auditory'. You notice what you hear, what is said, sounds, music and all aspects of what you hear in terms of volume, pitch, tone and pace. Sounds that may not be noticed by a visual person could set your teeth on edge and sights that alarm a visual person you may not even be aware of. You will hear someone talking on a mobile phone and probably find yourself listening in on the conversation because you can't block it out. You will notice squeaky doors, engine noises in the car, strange noises at night and hear the rumble of distant thunder before anyone else. You are automatically tuned into noise and sounds, both those you like and those you don't like.

You will be sensitive to the tone of the words someone says to you as much as the content, and you remember conversations word for word. If someone is talking too fast you will ask them to slow it down, and you yourself probably talk more slowly than your visual friends because as they talk fast, trying to express the images they see in their mind and translate them into words, you are talking slowly as you place importance in choosing the words you need to express your thoughts correctly. You enjoy conversations, both the talking and the listening, and prefer to talk to your friends rather than text or email them.

- **Auditory people tend to be quite still when they talk and rarely use gestures.**
- **Music is important to you. You may play an instrument or enjoy listening to music, probably both.**
- **When you are learning something new, you want to have it explained slowly and clearly rather than being given written instructions. If someone does that, you will probably say, 'Could you just talk me through this?'**

Here are some other words and phrases that are indicators of an auditory representation.

To tell you the truth..... We seem to be talking the same language I hear what you're saying What's the message here?

Exercise 11

Now it's your turn. Write some phrases in the speech bubbles that you might expect to hear from an auditory person.

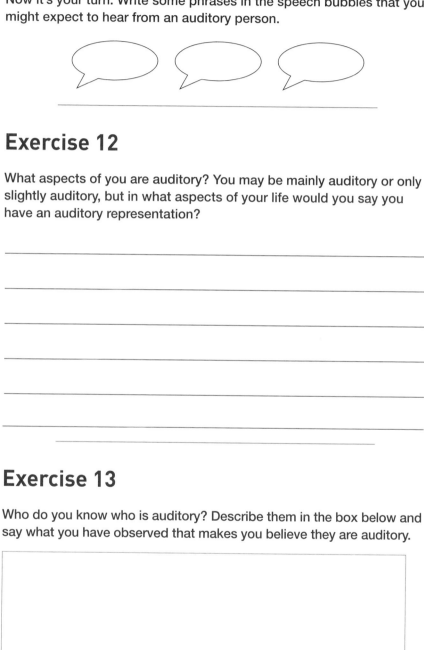

Exercise 12

What aspects of you are auditory? You may be mainly auditory or only slightly auditory, but in what aspects of your life would you say you have an auditory representation?

_____

_____

_____

_____

_____

_____

Exercise 13

Who do you know who is auditory? Describe them in the box below and say what you have observed that makes you believe they are auditory.

Kinaesthetic people

If you ticked mostly 'c' answers, then you have a kinaesthetic representation. You represent your thoughts as feelings, either internal feelings or external ones such as physical touch, taste or smell. You learn by doing and you prefer to be doing something rather than just watching or listening. For you, doing the exercises in this workbook will form an important part of your learning. You prefer to interact and take part. You like to connect physically with people when you are with them, possibly touching them and being physically close to them. You probably use gestures when you speak, using your hands to express yourself and moving around quite a bit. A kinaesthetic speaker will walk around and want to interact with his audience. Your speech will be quite slow, with pauses.

Kinaesthetic people can be fidgety, always on the move and often sporty. You will be sensitive to temperature and being comfortable is important to you, physically and emotionally. You will notice awkward atmospheres or tension between people and want to change it so that it is more comfortable.

Here are some of the phrases you will hear from a kinaesthetic person.

Get a grip!

Hold on to that thought.

Have you got that?

Let's get that in motion!

Exercise 14

Now it's your turn!

Exercise 15

What aspects of you are kinaesthetic? Can you give an example from your own experience when you show a kinaesthetic representation?

_____

_____

_____

_____

_____

_____

_____

Exercise 16

And what about friends and people you know? Who comes to mind when you think of people with a kinaesthetic representation?

_____

_____

_____

_____

_____

It is useful to know how you operate, so that you can recognize it and notice when you are matching (speaking in the same way as) the person you are talking to or mismatching (talking in a different way). It is then your choice whether you want to switch to their representation in order to achieve rapport and increased understanding, or whether you want to continue with your own preferred style. Hierarchy may dictate this. For example, you may want to match your boss or client because you want to be respectful and for them to listen to what you have to say. This will be more effective if you match them because they will recognize that you are 'talking their language'. In most cases, matching will be the most resourceful choice.

Exercise 17

Think about the person you would like to influence right now. What is their preferred style? Are they visual, auditory or kinaesthetic? Think of three ways you could match them in terms of language patterns, way of talking and body language.

I want to match a ....person.

	Words	Way of talking	Body language
1			
2			
3			

Exercise 18

This exercise is to test your versatility because, although you have a representation, there will be many occasions when you want to match someone else's.

Describe the last time you went out for a meal in three different ways.

→ Visual account

I saw _____

_____

➜ Auditory account

I heard _____

_____

➜ Kinaesthetic account

I felt_____

_____

Which did you find hardest? (tick which one)

➜ Visual ❑

➜ Auditory ❑

➜ Kinaesthetic ❑

This one will take practice, so notice when someone uses that representation and work hard to match it. Notice how the rapport builds as a result.

Each of these three representations has another dimension, internal and external, and when we work on strategies later this will become important, so let's get to know them now. They are:

- **visual internal – using images in our head to recall memories or construct ideas**
- **visual external – images we see with our eyes**
- **auditory internal – inner voice**
- **auditory external – what we say and hear**
- **kinaesthetic internal – our feelings**
- **kinaesthetic external – our actions**

Focus points

- You have now discovered the extent to which you prefer a visual, auditory or kinaesthetic process of communication. Are you V, A or K?
- You have been introduced to the type of words that signpost each representation, the body language and the way people speak.
- As you go about your daily life, interacting with people at work and socially, listen to someone speaking on the television or radio, notice the language they use and spot which representation – V, A

or K – they have. Can you spot it just by looking at them? Turn the volume off on the TV and see what you notice about the speaker's body language.

- Knowing whether you are visual, auditory or kinaesthetic helps you make choices in life: choices about how you want to learn, the type of job that will suit you, what type of holiday to book and who to date.

- Once you know which representation your partner has, you will also know how you can introduce them to a new idea, persuade them to do something they are reluctant to do and understand how their priorities in life may be different from yours.

 # Where to next?

In the next chapter, we learn about the meta-programmes of NLP. They are sometimes described as filters because they are ways of communicating filtered by our beliefs and values. We will cover these in more detail in Chapter 4.

We are not our filters; they operate all the time at an unconscious level. In the next chapter, we will be making them conscious so that we can recognize them in others, match them in others for building rapport and also choose to be curious about what consciously using a different filter would bring to our life. Once you read about them, you will instantly recognize them in others and start noticing the ones you use regularly. There is no judgement involved; there is no right or wrong filter, they are simply filters. You may have inherited them from parents as you copied their patterns as a child or you may have developed them as you developed your beliefs about how the world operates based on experiences from the past. Like beliefs, filters are not set in stone and if, when you find out what yours are, you choose to experiment with using another one, this is a choice you can make consciously.

Meta-programmes: more personal representations and what they mean

By being flexible about how we communicate, we have the best chance of connecting and influencing the result of the communication. Now that you know whether you are visual, auditory or kinaesthetic, you can use this knowledge to influence and build rapport with those you come into contact with who have other representations. In addition to the VAK, Bandler and Grinder also discovered that we run programmes in our thinking process which act as filters. They are not conscious processes but by understanding them we bring them to our conscious level in order to give us more flexibility. This chapter explains how by making choices in how you process what is happening around you can influence it to your advantage.

→ Understanding our filters

Every day we make millions of unconscious choices about what we pay attention to and what we don't. By being aware of the filter we are running and the alternatives available, we can change it and choose one that would be more resourceful. You might want to change it to match the person you are communicating with to improve understanding and rapport or change it to suit a particular task.

To understand more about ourselves and how we live our life, it helps to be aware of these filters, which are:

- towards/away from
- match/mismatch
- big chunk/small chunk

- choices/process
- internal/external

As we have seen, Grinder and Bandler found that what makes the difference in effective communication was not the experience itself but the meaning we put to it. One of the fundamental principles or ground rules of NLP is that the meaning is what one takes from the communication, not what was actually said. This is because we filter what is said according to our perception of the world. They developed what they called meta-programmes to describe the different filters they observed.

Each filter is named according to each end of a continuum; rather than you being at one or the other end, most people slide around a bit depending on the situation or who they are with although, like the VAK, there will be a tendency to be in one half or the other most of the time.

Exercise 19

Towards/away from

To find out which of these two filters you use, here's another questionnaire for you.

1 Why do you take out insurance?

a To have peace of mind ❑

b To avoid worry if something is taken or damaged ❑

2 What kind of goals do you set for yourself?

a Things I want ❑

b Things I don't want ❑

3 When you are playing a sport, what do you prefer to focus on?

a Winning ❑

b Not losing ❑

4 When you are applying for a job or a promotion, what do you aim for?

a Getting it ❑

b Not being turned down ❑

5 **When you meet someone new, what do you hope for?**

  a That they will like me ❏

  b That they won't dislike me ❏

I expect you can see straight away that the 'a' answers are all about aiming for a positive result in what you do and the 'b' answers are about avoiding a negative, which isn't the same thing at all. Do you know someone who is always saying 'but'? The chances are they are 'away from' because they have numerous reasons why they can't achieve whatever they want to do, whether that is finishing a task on time, enjoying their weekend or getting that great new job.

One of the classic 'away from' goals we often set ourselves is 'losing weight'. Imagine that: we are focusing on what we don't want, i.e. weight. Wouldn't it be better to focus on what we want more of? Perhaps we want more looseness in our clothing around the waist, more admiring glances, more spring in our step, more energy? All these would be 'towards' goals. Grinder and Bandler found in their research that successful people had 'towards' goals. So if you want a nice trim body, focus on visualizing yourself as slim rather than 'not fat' and you'll be more likely to succeed.

Exercise 20

Think about a goal you have for yourself in this exercise. In the space below, write down what you want to achieve by the end of this workbook.

_____

_____

_____

_____

Read what you've just written. Have you worded it as a 'towards' goal? Check for any negative words – any 'buts', 'don'ts', 'avoid' or 'not'. If you have used any, reword your goal in the next space so that it is a compelling vision that you really want for yourself.

_____

_____

It is like the idea of the carrot and the stick, where 'towards' thinking is like being the donkey motivated to move forward by the carrot dangling in front of it and 'away from' thinking is the stick behind it.

Exercise 21

Match/mismatch

Find out which of these two filters you use, with this questionnaire.

1 **When you meet someone new, what do you notice more?**

  a What you have in common ❑

  b How you are different from them ❑

2 **In conversation, do you tend to agree or disagree with the other person?**

  a Usually find areas in which you agree ❑

  b Tend to find things you disagree with ❑

3 **Do you follow fashion or do your own thing?**

  a Follow fashion ❑

  b Adopt my own style ❑

4 **If your friends follow a particular soap or TV programme, do you?**

  a Watch it too ❑

  b Watch something else to be different ❑

5 **Do you enjoy a good argument?**

  a No ❑

  b Yes ❑

If your answers were mostly 'a', you prefer to be in agreement and look for similarities, whereas mostly 'b' answers indicate that you prefer to look for what is different. There is no right or wrong way. People who are good at finding the mismatch can be excellent problem solvers, fixers of things and good at investigating anomalies in systems. People who have a tendency to match are good at getting on with people and building rapport and supportive relationships, such as in customer service.

People who are good at matching find a connection in their own experience with what the other person is saying and tend to respond with 'Yes, and...', adding in some way to what they are being offered. Of course, if this is forced it can appear insincere. Mismatchers often look for a flaw in what the other person is saying, by saying, 'Yes, but...' and this can appear critical although, done with good intention, it can encourage new ways of thinking.

Matching and mismatching can be done without even speaking. You can build rapport by adopting the same body language as the person you're with, matching their volume, pace and tone, laughing with them and returning their smile. If you want to end an interaction, you could deliberately mismatch to lose rapport.

The ideal is to have the flexibility to be either, depending on the outcome you want from the interaction or task. As you'll remember from the introduction, the person with the most flexibility controls the system. Let's see how flexible you are!

Exercise 22

Think of a situation in your life where matching would be the most appropriate strategy.

_____

_____

Now think of another situation when it would be appropriate to mismatch.

_____

_____

Exercise 23

Thinking of your friends and family, work colleagues and those you interact with on a regular basis, divide them into two groups – matchers and mismatchers.

Matchers ('yes and')	Mismatchers ('yes but')

Which group do you find it easier to be with?

_____

Exercise 24

Big chunk/small chunk

Find out which of these two filters you use, with this questionnaire.

1 Thinking ahead to the weekend, do you have a rough idea of what you will do or do you have everything planned in detail?

  a Just a rough idea ☐

  b Everything planned in detail ☐

2 If I ask you about your job, will you give me a brief overview or a detailed account of what you do?

  a An overview ☐

  b Detailed account ☐

3 **When you prepare a new recipe, do you follow it in every way or use it as a guide?**

 a Use the recipe as a guide ❏

 b Follow every step carefully ❏

4 **Do you read the small print of contracts and instructions?**

 a Just a quick skim through to get the main points ❏

 b Read it all ❏

5 **Are you good at organizing or better at having the ideas?**

 a Great at generating ideas but less good at organizing ❏

 b Better at organizing if someone else has the ideas ❏

If you are big chunk you will have chosen mostly 'a' answers. If you are small chunk you will have chosen mostly 'b' answers. There could be a mix of answers if you look for detail (small chunk) in some situations but not others. For example, if you are not a competent cook you will need to follow a recipe closely, and if you are buying a house you would read the small print for such an important purchase, even if you don't normally do so. This is more about what comes naturally to you.

Sometimes we need to 'chunk up' or 'chunk down'. What this means is that if we are getting caught up in the detail of something and cannot see the wood for the trees, we need to take a step back and ask ourselves what it all means, so that we can see the bigger picture. Similarly, sometimes when we cannot see our way forward and can only see the big picture – for example, 'I want a new job' – then we need to chunk down and write down what the small steps are towards the big-picture goal. You can do this chunking with someone when they are getting bogged down in detail or when they are not being specific or detailed enough for what you need.

Let's have a go at this.

Exercise 25

Chunking down and up

A big-chunk task is doing the shopping. What are the small-chunk aspects of this task? Write them down in a sequence step by step.

1 _____

2 _____

3 _____

4 _____

5 _____

6 _____

7 _____

8 _____

9 _____

10 _____

Now here are some small-chunk tasks. Write in alongside what the big-chunk task is.

→ Reading the map

Because I need to _____

→ Peeling some fruit

Because I want to _____

→ Turning on my PC

Because I'm going to _____

→ Phoning a friend

Because I want to _____

→ Going to a conference

Because I want to _____

Become fluid at recognizing when you need to chunk up or chunk down and when you can help others by doing this.

Exercise 26

Choices/process

Do this questionnaire to learn whether you prefer options or procedures.

1 Imagine you want to buy a new car. Will you enjoy considering all the options?

   a Yes, that's the best part of the process ❑

   b No, I just want to get the decision made ❑

2 In a restaurant, would you choose something like a mezze or dim sum or a specific dish?

   a Enjoy smorgasbord, dim sum or mezze ❑

   b A specific dish ❑

3 Does the idea of being freelance appeal?

   a Yes, I want to be in control of my destiny ❑

   b No, it's pretty scary; I'd rather be employed ❑

4 Are you good at starting new projects?

   a Starting them, yes, but not so good at seeing them through to the end ❑

   b No, I prefer someone else to start them and I will see them through to completion ❑

5 **How would you make a journey you do frequently?**

   a Vary the route so I don't get bored ☐

   b Always take the same route ☐

If you like to run the choices – or options – filter you will have mostly chosen 'a'. This is because you enjoy finding new ways of doing things, ways of bending the rules and a variety of ways to get the outcome you want. You are creative and are probably known and admired for your ability to come up with new ideas, but carrying them through to completion may not be one of your strengths.

If you have chosen 'b' this is because you run a more logical filter and like to do things in stages, step by step, following a set of rules and procedures. You may often find the idea of choice overwhelming.

Remember that filters or meta-programmes are like a sliding scale: in this case from options to procedures, from choice to a step-by-step process. Undoubtedly, you'd like to operate in one part of the scale for some tasks or experiences and in another part for a different one. There is no right or wrong place to be on the scale, but it is a matter of adapting to what is needed to complete a task or communication in a resourceful and effective manner.

Exercise 27

Internal/external

Find out from this questionnaire whether you are internally or externally referenced.

1 **If someone at work criticizes you, how do you respond?**

   a Take on board whatever they say; they must have a point ☐

   b Think about what they've said and decide whether I agree that they have a point ☐

2 **What type of clothes do you prefer to wear?**

   a What everyone else is wearing of course; I don't want to stand out ☐

   b Styles that reflect my own personality and tastes ☐

3 **Before going to a new movie, what do you do?**

  a Read the reviews and ask people who've seen it, and then decide
     if I want to go ❏

  b Decide for myself whether or not to see the movie based on my
     own criteria ❏

4 **Someone offers you some chocolate but you're on a diet. What do
  you do?**

  a Take it because I don't want to offend them ❏

  b Refuse politely and explain why ❏

The 'a' answers suggest you are externally referenced, which means that
what other people think and say matters more to you than what you think
or say. You are influenced heavily by others and want to be liked and
to fit in with the crowd. If you gave more 'b' answers, you are internally
referenced and rely on your own judgement without reference to others.
You check in with your own values and beliefs before making decisions,
regardless of the circumstances.

Bandler and Grinder found that by matching your meta-programme
with the person you are speaking to, you can best influence the outcome
and build rapport.

Exercise 28

In this exercise you must respond to each speaker using their
filter. It does not matter what you say as long as you match their
meta-programme.

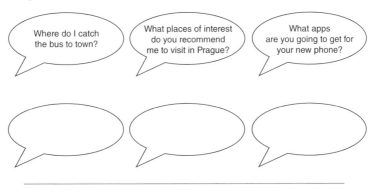

Where do I catch the bus to town?

What places of interest do you recommend me to visit in Prague?

What apps are you going to get for your new phone?

Exercise 29

Choosing your filter

Are you options or procedures? Let's have a go at each one and see how flexible you can be.

→ Describe your last holiday as if you were running an options filter.

_____

_____

_____

_____

_____

_____

→ Now describe it running a procedures filter.

_____

_____

_____

_____

_____

_____

_____

Exercise 30

Detecting filters

To see whether you can now detect all these filters, write down the filter you detect for each of the following statements:

→ I just can't decide what colour to paint this room. I could go for a bland colour, make a statement or have a feature wall. I'm enjoying looking through all the paint charts.

_____

_____

→ I hear what you're saying but I don't agree. I think we should reduce the price of Product A.

_____

_____

→ I generally find I buy the same brands as my friends.

_____

_____

→ Whatever you say, you're usually right about these things.

_____

_____

→ I hope I don't get seasick on the crossing.

_____

_____

→ I always follow the instructions when I put together flat-pack furniture.

_____

_____

→ It's important to learn a language step by step.

_____

_____

→ I'm going to be size 10 by Christmas.

_____

_____

→ I tend to do my own thing.

_____

_____

→ I generally follow my nose when it comes to finding
my way around.

_____

_____

Remember that it is the person with the most flexibility who influences the system. There will always be people with different filters who you need to get on with at work or at home, and tasks that require a specific filter. You also need to be able to recognize when you can get a more positive outcome by changing your filter.

Focus points

- By completing the quizzes, you have learned where on the continuum of each meta-programme you tend to be, although this can vary by situation, by task and even who you're with.
- If you have a mismatch filter you may find yourself operating the opposite filter from the person you're with, just to mismatch!
- Past experiences and memories can also affect our filters. For example, a normally 'towards' person may be 'away from' in connection with a specific health issue if it runs in the family.
- Grief and depression can also affect our filters; this is called our 'state'. When in a particular 'state', a person may adopt a different filter. For example, when in grief we may find 'choices' too difficult to cope with and prefer to run a process filter just to get through each day.
- Filters or meta-programmes are like a sliding scale: there is no right or wrong place to be on the scale but it is a matter of adapting to what is needed to complete a task or communication in a resourceful and effective manner.

Where to next?

Behind the filters we unconsciously use is a set of beliefs and values that underpin our identity and are themselves underpinned by our behaviour and skills. These are all part of what NLP calls the logical levels of change, which we will explore in the next chapter. Beliefs and values passed on in families will ensure that filters are passed on with them. For example, a mother who constantly wraps up her child 'against the cold so they don't get ill' warns them against taking risks or overdoing it, and has herself avoided anything that might be stressful, will pass on an 'away from' filter to a child who then grows up to take cold prevention remedies, avoid stress and lack drive because they have a belief that 'life can be dangerous so we have to guard against taking chances'.

Knowing the filters you use will enable you to explore where they came from and how they affect the choices you make in life. This gives you the opportunity to change them and give yourself more resourceful choices. NLP is full of choices, making unconscious ones like filters, conscious.

4 Chunking up and down the logical levels

One of the presuppositions of NLP is 'the map is not the territory', meaning that whatever we are and whatever we do is within a wider context and affects others. Each person's map is unique, and it is useful to explore this map through what NLP calls the 'logical levels of change', which range from environment at the lowest level through behaviour, capabilities and skills, values and beliefs, to identity and purpose at the highest level. Through the exercises in this chapter, you can explore each level before moving to the next, and this will give you a road map for change.

→ The logical levels pyramid

Robert Dilts devised the logical or neurological levels pyramid shown overleaf. The levels are presented as a pyramid because each level is based on the one below it and metaphorically points upwards towards your purpose in life. It is rather like Maslow's well-known 'hierarchy of needs' pyramid, which starts with the basic physiological needs of food and shelter at the base and ends with self-actualization at the top, which is the highest level of need and similar in concept to purpose in the logical levels.

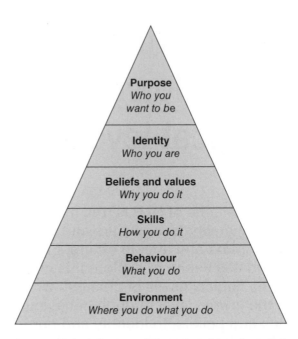

We are going to start at the top of the pyramid and work down, but equally you can start from the bottom or even in the middle. Wherever you start, the important thing is to move on to the adjacent level and to ensure you look at every level in turn. This is because each level affects the level next to it; they are interrelated and connected. The reason we are starting at the top is to look at the goal you have in mind, your objective for reading this book, something positive that you want to change to or for.

Purpose

At the top of the pyramid is **purpose**. This is like your own personal mission statement. What you want to achieve is bigger than who you are, it is who you want to be. What do you want from life? What are you here on earth to do? What would you like to be remembered for? What would you like people to say about you?

These are quite big questions – certainly big chunk – so if you generally run smaller chunk filters (see Chapter 3) you will need to chunk up. Remember that we do this by writing something down and then asking ourselves 'What does this mean?' or 'What would this look like or sound like or feel like at the next level?'

These are just some of the questions to ask yourself in order to write down in the space below what your purpose is. Who or what do you want to be?

_____

_____

_____

_____

We will be looking at your goals more thoroughly in Chapter 6, so treat this chapter as a way to prepare yourself for this. If your goals change, you can return to the logical levels, knowing how they work.

Identity

The next level down is **identity**. This concerns what sort of person you are. This is not what you do for a living or how you spend your time but _who_ you are. Many of us are in jobs that pay a salary and are not too demanding so we have time to be what we want to be outside work.

> # Example
>
> Edward has always worked for companies that allow him to work 9–5 and rarely require him to travel or work late. The work is not too stressful or demanding intellectually, so he is able to cycle to and from the station and do regular long cycle rides at the weekend. Edward's identity is that he is a cyclist but this is not nor ever will be his full-time occupation. What he eats is defined by being a cyclist, where he goes on holiday has to allow him to cycle and he would never live somewhere that did not offer good cycling terrain, nor would he work abroad somewhere where he could not cycle.

What is your identity? Who are you when you are being you? If you had to sum yourself up in three words that would define you – so that you would recognize yourself and those close to you would know it was you – what would they be?

I am a _____

I am a _____

I am a _____

If you've never thought about this before, what you've just written could be quite powerful. Before you go any further, just record how you feel about making this declaration about your identity.

_____

_____

_____

_____

_____

_____

_____

_____

Now let's delve some more into your identity.

Exercise 31

List in the following spaces the adjectives that define you. Choose the adjectives that you would use to describe yourself, that others have used and ones that are important to you. (Imagine if someone said that you were not, say, creative, you would be horrified, because it is your belief that you are that adjective and it matters to you to be so.)

1 _____

2 _____

3 _____

4 _____

5 _____

6 _____

7 _____

8 _____

9 _____

10 _____

Look at the adjectives you have listed and underline those that are absolutely fundamental to who you are and write them into sentences below, saying why this is your belief.

I am _____

_____

_____

I am _____

_____

_____

I am _____

_____

_____

I am _____

_____

_____

I am _____

_____

_____

Just think now for a moment whether what you are experiencing right now is an expression of who you are. Sometimes in life we find ourselves driven by events, our job, our home life and so on, into having to take on an identity that does not fit us and where we do not feel comfortable. Are you in that place? Hold on to your identity as you travel down the logical levels.

_____

Beliefs and values

The next level, **beliefs and values**, concerns what you hold dear and the way you live your life. They are the 'whys' about what you do. Beliefs are the ideas that you hold to be true at the moment. Beliefs change; after all, perhaps you believed in Father Christmas or the Tooth Fairy as a child but probably don't any more! Our beliefs change all the time as we experience new things, meet new people and travel to different cultures.

Exercise 32

Complete the following sentences to help you discover your beliefs and values.

What gets me out of bed in the morning is the belief that

_____

_____

People who work hard at their job should

_____

_____

People who commit a terrible crime should be

_____

_____

Teenagers should be prevented from

_____

_____

A good mother is one who always

_____

_____

Old people should

_____

_____

Children must be

_____

_____

A good teacher is one who

_____

_____

My ideal boss is one who

_____

_____

A good movie is one that

_____

_____

People should be treated as

_____

_____

When you travel to another culture you should

_____

_____

The government should not

_____

_____

I do not believe in

_____

_____

If you want to get better at something you have to

_____

_____

The best-paid jobs are in

_____

_____

_____ should be paid more.

_____

_____ should be paid less.

_____

Companies must not _____

_____

It is important to be _____ at all times.

_____

Are there any surprises here? Have your answers sparked off other thoughts about your beliefs? Write them down in the space below so that you can refer to them later.

_____

_____

_____

_____

_____

Values are the things that are important to you, your code for life. They might include things like honesty, trust, peace and so on.

Exercise 33

Here are some more questions to answer. They won't all be relevant, so just answer those that are.

What I value about my boss is the way he/she

_____

_____

What I value about my relationship is

_____

_____

What I value about my team at work is

_____

_____

What I value about my child is

_____

_____

What I value about my friend is how he/she

_____

_____

The reason I applied for this job was because I valued

_____

_____

The reason I play sport is because I value

_____

_____

The reason I save is because I value

_____

_____

My friends would say I value

_____

_____

My family would say I value

_____

_____

_____

Exercise 34

Now place your answers from the previous exercise in order of importance to you.

Rank 1 is _____

_____

Rank 2 is _____

_____

Rank 3 is _____

_____

Rank 4 is _____

_____

Rank 5 is _____

_____

Rank 6 is _____

_____

You may find, having done these exercises, that there is a mismatch between what you are currently doing in life and your values and beliefs. If this is the case for you, think, as you continue on down the logical levels, about what you might want to change in your life in order to get your values and beliefs aligned all the way through the levels. Alignment enables us to feel comfortable and confident in our skin.

To summarize this section on beliefs and values, complete these sentences.

I do what I do because I value

_____

_____

_____

_____

_____

I do what I do because I believe

_____

_____

_____

_____

_____

_____

Skills

The next level down is **skills**. Skills are the things you do well. Because you do them well you may not even be conscious of some of them as skills. Some of your skills will be those you have specifically learned, perhaps for work or when you learn a new language, perfect a recipe for a dinner party or learn a sport. These will be conscious skills. Other skills you will have learned from parents and friends, or picked up from others almost organically, and you'll only be aware of them when someone comments on them and brings them to your conscious level.

Example

Rebecca was asked what she could do well. Her mind went blank. She thought of all the things she could do but she assumed everyone could do them and probably better than she could. She had low self-esteem and practically mumbled, 'Well, I can do a cartwheel.' She was astounded when everyone in her group looked amazed and said they couldn't do one and would love to learn how.

She took them all outside and showed them. Their attempts were hilarious and she confirmed that indeed she was the only one there who could do a cartwheel. Aha, she thought, I do have a skill after all. Then she started thinking of other things she could do and there was quite a list.

Exercise 35

This exercise will help you identify your skills.

Does the idea of entering a talent contest terrify you? What would you offer? Without giving it too much thought, write three things you could do. Remember that not everyone can do what you can do.

Well, that wasn't so difficult, was it?

 # Exercise 36

Friends often compliment us on what they notice that we do well. Think back. What do your friends say you do well? If someone were to ask them about what you are good at (out of your earshot), what would they say?

Take these on board and think about them. Your friends notice things about you that you don't notice because perhaps they are things you've always done and they come naturally to you. They are done unconsciously. It's the same with your family. What do they say you do well?

What about your partner; what would he/she say that you do well?

Now you have a long list of skills. We haven't finished with them yet. Look at each one and think about how you came to do it well. Are there common threads or themes? How do they relate to your VAK (visual, auditory, kinaesthetic) preference and your meta-programmes (filters)? What have you learned from these last few exercises around your skills? Write down what you've learned in the box below.

Behaviour

The next level down is about what you say and do: your **behaviour**. This is the outward expression of your skills, beliefs and values, identity and purpose. Is your behaviour truly reflecting these at the moment? Your behaviour is the actions you take, what you do on a daily basis. Is what you do aligned to your skills and capabilities, your beliefs and values, your identity and ultimately your sense of purpose?

It is not unusual to find that, as a result of circumstances, we are not doing what makes us aligned. Perhaps we found ourselves in a job that paid well and we progressed in our career without realizing that our skills and values may be more suited to something different. Ask yourself now whether this could be the case for you.

Exercise 37

What do you do on a daily basis? What functions and tasks do you perform? List them below.

1 _____

_____ ❏

2 _____

_____ ❏

3 _____

_____ ❏

4 _____

_____ ❏

5 _____

_____ ❏

6 _____

_____ ❏

7 _____

_____ ❏

8 _____

_____ ❏

9 _____

_____ ❏

10 _____

_____ ❏

Now go back and put a tick alongside those that you feel fit with who you are and the skills you have. How many are ticked? ❏

What could you change at the behavioural level that would give you alignment?

→ What could you do more of?

_____

_____

_____

_____

_____

_____

→ What could you do less of?

_____

_____

_____

_____

_____

If there are things you do that you don't want to do, you may need to look at making a change at the identity level and then at the skill level. In order to access the skill, you need to change the behaviour to meet the identity change. Ask yourself, 'What do I need to do at the behavioural level in order to be who I want to be at the identity level and what skill do I need for that?' Then you can decide whether that fits with your beliefs and values. If it doesn't, you can change it.

Look back at those behaviours that are not ticked. Why do you do them? Check whether they fit at another level. Perhaps they fit with your values and beliefs? Do they fit with your identity? Is there another way to meet those needs using a different behaviour that would better match your skills? Use the space below to note what behavioural changes you could make towards alignment.

_____

_____

_____

_____

_____

Environment

Your environment is where, when and with whom you do what you do. This is about the time we live in, where we live and the people we interact with. Consider your environment today as it is now.

 ## Exercise 38

Thinking about where you spend your time and with whom, what is working well for you and what could you change so that you could be more aligned?

These things are working well	These things could be improved

What changes could you make in your environment that would enable you to become more aligned?

_____

_____

_____

_____

_____

_____

Finally here's the logical levels pyramid again. Having done all the exercises and read about the different levels, take yourself up the levels now with the changes you have made in place and check that you are now aligned. If you want to change at one level (or want to help someone else to change), you need to make a change first at the level above where you want the change to occur.

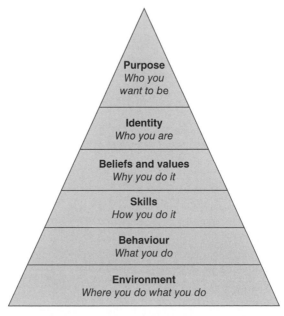

Focus points

- The logical or neurological levels of change are presented as a pyramid because each level is based on the one below it and metaphorically points upwards towards your purpose in life.
- Each level in the pyramid impacts on the level adjacent to it, above and below.
- As we seek to make changes in our life in order to become aligned through the levels, we have to start exploring at a level above the one where the change needs to occur.
- Keeping a constant check on your levels in order to be aligned is an ongoing process; with each change or decision we make in life, we need to check for alignment.

Questions to ask are:

- Does this fit my purpose in life?
- Does this fit with who I am as a person?
- Does this fit with my beliefs and values?
- Do I have the skills to do this or can I learn them?
- Can I do it?
- Where, when and with whom can I do it?

 # Where to next?

In the next chapter, you will learn about rapport, which is about getting on with people by building a mutually respectful relationship built on trust and understanding. It will build on all you have learned in the chapters on VAK and the meta-programmes in terms of how we communicate and process information. It follows this chapter because we will be successful at rapport and in building relationships if we ourselves are already aligned.

5 Getting on with people

An NLP principle is that behind every interaction is a positive intent. We want to get on with people, be in rapport, so how can we achieve this, even with people who process their world very differently from us? How we interact with others is programmed by our beliefs and values, identity, skills and all the factors covered in Chapter 4 on the logical levels, and filtered by our VAK and meta-programmes, which by now will also be familiar to you.

→ Finding rapport

Rapport is the word we use to describe the way we interact with others. Good rapport is when we feel at one with the other person, we understand each other and enjoy each other's company. Bad rapport or lack of rapport is the opposite, and occurs when we feel uncomfortable, not understood and not liked. When rapport is good we feel good inside, but may not realize why. Similarly, with lack of rapport we may feel uneasy without knowing why. Since we can therefore be unconscious about rapport, by bringing it to consciousness we will be able to ensure that when we need to have rapport, we know how to achieve it.

In terms of rapport, you are the message. What someone takes from you, regardless of your intention, is the message. This means that if your body language does not match what you are saying or if your voice tone doesn't match, the message will be distorted.

Did you know that 93 per cent of communication is through body language and tone of voice and only 7 per cent is down to the words you use? You can experiment in the following exercise.

Exercise 39

Read the following sentence in your normal voice, standing or sitting naturally as you are doing now.

'I don't understand what you're saying.'

Now say it again in the following ways.

- Slowly
- Fast
- Low voice
- High voice
- Hands on hips
- Hand under chin
- Head to one side
- Emphasize 'I'
- Emphasize 'don't'
- Emphasize 'understand'
- Emphasize 'what'
- Emphasize 'you're'
- Emphasize 'saying'
- Smile
- Frown
- Look away
- Look down
- Look up

We could go on, but the point has been made, hasn't it? The same words can take on a completely different meaning, and rapport depends on how they are delivered. In this age of global business when so much is communicated by email and text, there is a danger that we lose some of the social skills of communicating face to face. We may inadvertently communicate quite a different message from the one intended.

We tend to socialize with people just like us, with whom we already have rapport and shared history, but in life we also have to get on with people who aren't like us, so in those cases we have to work at it. Remember that flexibility is a key component of NLP and it is by being able to be flexible, using our knowledge of VAK and meta-programmes, that we can adapt seamlessly to others who have a different way of processing and whose different life experiences mean that their values and beliefs are different.

In addition to being flexible, good rapport comes down to being curious about the other person's map of the world, which may be different from yours. By keeping an open mind and being curious about their positive intention we can build rapport, even with people with whom – on the face of it – we have little in common.

Matching

Sometimes we need rapport in order to influence others: our children, employees, members of our team at work or members of a sports team, people on a committee and so on. By using the NLP skills explained in this book, we will learn to influence with integrity through matching. Matching is choosing to match the other person's internal representation and meta-programmes so that they recognize that we are 'like them'; we speak the same language and share their beliefs and values.

Chapters 2 and 3 focused on understanding VAK and the meta-programmes. Through listening out for how people think, we can formulate what we say so that they make a connection with us. We know now that if we match the person we are talking to, both in terms of their body language and how they speak, we will make a good connection and be well placed to build rapport.

 # Exercise 40

 If you look at two people who are in rapport, what do you notice?

What do you see?	What do you hear?	What do you feel?

Exercise 41

Is there someone you'd like to get along better with? Think first about why you want this and what your desirable outcome is. Write it down here in the space below.

_____

_____

Think about them and write their name here.

_____

_____

Let's consider the areas we need to work on by reminding ourselves of our VAK and meta-programmes.

→ My preference is **visual/auditory/kinaesthetic** (circle as appropriate).

→ She/he is **visual/auditory/kinaesthetic** (circle as appropriate).

How could you alter the words you use so that you communicate in a way they can understand? Think about using words that fit their preference. Write them down in the space below.

_____

_____

_____

_____

_____

Think about the meta-programmes and, again, circle the one that applies.

→ I tend to be **towards/away from.**

→ She/he tends to be **towards/away from.**

If they are different, how could you express yourself to match their preference?

_____

_____

_____

_____

_____

→ I tend to be **match/mismatch.**

→ She/he tends to be **match/mismatch.**

This can often be an area of conflict because if one of you likes to be in agreement and the other likes to find a flaw in what is said, this can feel quite uncomfortable. How can you match with their preference?

_____

_____

_____

_____

_____

→ I tend to be **big chunk/small chunk.**

→ She/he tends to be **big chunk/small chunk.**

If one of you likes detail and the other avoids it you can find it difficult to communicate. How can you chunk up or chunk down to match them?

_____

_____

_____

_____

_____

_____

→ I tend to be **choices/process.**

→ She/he tends to be **choices/process.**

If one of you likes choices and the other prefers a step-by-step approach without choices to distract, this can lead to frustration and miscommunication. How could you improve the communication between you using this knowledge?

_____

_____

_____

_____

_____

_____

→ I tend to be **internally/externally referenced.**

→ She/he tends to be **internally/externally referenced.**

If this is different you may need to connect with their referencing and reflect that back so they can relate to what you're saying. How could you do that?

_____

_____

_____

_____

_____

_____

Overall, how could you use what you've learned about VAK and the meta-programmes to improve that connection between you?

_____

_____

_____

_____

_____

_____

Sometimes not saying anything can be a really effective way to communicate! Just by tilting your head to suggest a curious attitude can encourage the other person to rephrase what they are saying in a way that might connect.

If there is some confusion or miscommunication between you, there are ways to ask for clarification without losing rapport. For example, saying 'What do you mean?' can sound confrontational, whereas 'In what way……(using their word)?' or 'How do you mean………..(using their word)?' sounds more curious and stays in rapport.

Another way is to repeat what they have just said but finishing on an upward inflection, indicating a questioning or curious response.

These types of responses are called 'clean' because they are devoid of assumptions. By reflecting back what we are given without adding any of our own 'stuff' the interaction stays in rapport and builds understanding. We are not trying to solve their problem or disagree but showing our desire to clarify. It is based on the NLP principle that every action has a positive intention. The other person's positive intention is to communicate, so we need to assist them in whatever way we can. A clean intervention means that they have the freedom to continue so that there is mutual understanding and respect and rapport is maintained.

The meta-model

There are three other aspects of communication that we need to understand in order to achieve and build rapport. Meta-programmes are one type of filter through which external events are processed and there are three others, which are called the meta-model. These are **deletions**, **distortions** and **generalizations** and they interfere with clean communication in various ways. Interestingly, they are also part of the Milton Model, when the same patterns are used deliberately to create a path to the unconscious by being deliberately vague. The Milton Model is discussed in Chapter 11.

Deletions

Deletions are when we are being vague and deleting detail from our communication so that the other person can't understand what we mean, or misses the communication completely. The listener has to fill in the missing information with his or her own assumptions, which may be incorrect. Here are some classic examples of deletions you will recognize.

Type of deletion	Examples	What's deleted
Comparison	'That was a much better presentation.' 'Your performance was worse.'	The detail, so we have no way of knowing what to learn from the feedback and what precisely we've done better or worse and in what way it was better or worse.
Judgement	'This is wrong.' 'You are wrong.'	The fact that this is the speaker's opinion. It is cleaner to say, 'I think this is wrong.'
Nominalization	'Feedback is difficult to take.'	What aspects of feedback are meant and the fact that, again, it is an opinion.
Vagueness	'She's annoying me.'	How she is being annoying.
Simple deletion	'I'm leaving.'	Information about why, where and how long for.

You know when information has been deleted because you want to ask a question afterwards, such as 'Who?', 'What?', 'When?', 'Where?' or 'How?'

Exercise 42

Can you say what is deleted in the following sentences?

➜ There he goes again! _____

➜ What shall we do? _____

➜ That's better. _____

➜ That's ridiculous. _____

➜ Arguments are upsetting. _____

➜ She's gone. _____

Another example of a deletion is when someone compliments us and we dismiss it with a shrug, saying, 'Oh, it's nothing.' Here, we are deleting the feedback. Are you someone who deletes compliments?

_____

Generalizations

Generalizations are when we base our expectations of one experience on all experiences. For example, you may say, 'I can't speak in public' because perhaps in the past you had a bad experience of public speaking and have generalized that this means you can't do it. Generalizations limit us and are easily recognized by words like 'always', 'never', 'should', 'must', 'ought to' and 'everyone'.

They limit us because they reduce our choice to be flexible, do things differently next time and discount possibilities to be different. They are based on our beliefs but, because they limit us, we call them **limiting beliefs**.

Exercise 43

Let's look at your limiting beliefs based on generalizations.

I always _____

_____

_____

→ If you didn't have this belief, what could you do?

_____

_____

_____

I should _____

_____

_____

→ If you didn't have this belief, what could you do?

_____

_____

_____

People like me ought to_____

_____

_____

→ If you didn't have this belief, what could you do?

_____

_____

_____

When I'm at work I never _____

_____

_____

→ If you didn't have this belief, what could you do?

_____

_____

_____

I can't bear to _____

_____

_____

→ If you didn't have this belief, what could you do?

_____

_____

_____

Next time you hear yourself saying things like this, ask yourself, 'And what if I could?' and see where that takes you. Question whether the limiting belief is helpful for you now. Maybe it's a belief you had when you were younger and could be revisited? You have choice and you can choose to change these beliefs.

_____

Distortions

Distortions occur when you make an assumption about what someone else is feeling or how they will react, or about what will happen next, without checking it out. We call it mind reading. The most common way to identify these distortions is by the expression 'made me'; for example, 'He made me lose my temper.' The fact is that no one *made* you do anything; you *chose* to respond in that way. When we assume we know the other person's intention or feelings, this is distorting their experience and basing it only on our own perceptions of reality. This hampers rapport.

Here are some examples.

She makes me feel stupid.

You must be so proud.

Don't make me laugh!

He knows what I mean.

When you hear yourself saying something that sounds a bit like mind reading, ask yourself, 'How do I know this?' or 'How did I make myself think this?' because otherwise you will find yourself losing rapport by distorting the communication.

Focus points

- Matching the internal representations and meta-programmes of others is key to achieving good rapport which will ensure that communication is understood and respected. Nearly all – 93 per cent – of our communication is non-verbal.

- Here are 10 quick ways to get into rapport and 'make friends':
  1 Adopt a similar body posture – it may seem strange to you but it will appear perfectly natural to them.
  2 Listen to them and pick out some of the key words and phrases they use and use them too.
  3 Use their terminology.
  4 Use similar hand gestures.
  5 Recognize their personal space: if they want to stand close, stay close, and if they want more space, keep apart.
  6 Speak at the same volume, pitch and pace.
  7 Avoid deletions, distortions and generalizations.
  8 Be curious and ask for more information, using clean questions.
  9 Match their chunk size.
  10 Match their VAK preference.

Where to next?

In the next chapter, we will learn about setting a compelling outcome, a goal. You may associate goals with work, and indeed they have a place in that context, but in NLP we learn that compelling outcomes are appropriate to every situation, even down to just a conversation or an email. When you think in those terms – that every action needs a compelling outcome – you will automatically know how to implement the NLP tools and techniques you have learned.

6 Goal setting the NLP way

Whether we want to set life-changing career or relationship goals, reduce our handicap in golf or learn a new language, it is important in NLP to create a 'compelling outcome'. In this chapter, we will look at doing this in exercise form in three different areas of our life – work, relationships and play/sport. We will use a timeline.

Whatever we want in life has a goal attached, whether it is simply a conversation with the boss or our intention to become fitter and healthier this year. In Chapter 3 we looked at goal setting in the context of making 'towards' goals, and you had the opportunity to frame your goal in that way. In this chapter, we will explore goal setting in more depth.

→ Creating a compelling outcome

In NLP we talk about creating a 'compelling outcome'. This is an important concept because it is the expression of what you want to achieve for *you*, what the result will be and what this will mean to you. A good way of looking at it is to ask ourselves 'How will I know when I have achieved it?' and, at the VAK level, 'What will I see, hear and feel that will tell me that I have achieved it?' You may want to look back at the logical levels in Chapter 4, where you wrote in your purpose at the top of the pyramid.

In business we are encouraged to consider whether our goals are SMART, that is:

- specific
- measurable
- achievable
- realistic
- time bound

What these factors lack is passion. How likely are we to achieve our goal if we do not have passion and 100 per cent belief? One only has to consider some of the great achievements of intrepid explorers, innovative leaders in technology, great sportsmen and women and passionate entrepreneurs to know instinctively that they would not necessarily have applied SMART goals because they had belief and passion beyond what was realistic or achievable.

We are going to use a timeline. First, think about your goal or compelling outcome and write it down here. Be bold: this is what you really, really want, remember, for you and for no one else.

_____

_____

_____

Remember to phrase it in positive terms and check that it is within your control because you can't control the decisions others might have to take for you to achieve your goal. You also cannot control winning a race, becoming the managing director of your organization, making your partner happy or, indeed, anything that relies on the performance of others or on external circumstances.

If you need to, rewrite your goal here. It will probably start with, 'I want to be...'

_____

_____

_____

Now we're ready for the timeline. We are going to do this exercise three times, for work, sport and relationships. Work covers what you do most of the time. It is probably how you earn your living or how you develop yourself, although it doesn't have to be.

 ## Exercise 44

Work

Find a space where you can draw an imaginary line along the floor. It needs to be about 2m (6ft) long.

Let's assume that one end of your line represents your childhood and the other end is old age. Now stand on the place that for you is today, the age you are or where you are today in a work context.

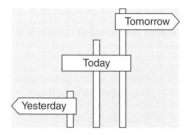

Remember your internal representation? Are you visual, auditory or kinaesthetic?

As you stand there, think about how your working life is for you now. You may want to close your eyes to block out visual distractions or, if you are auditory, turn off any music and close the door so you can focus. Here are some specific questions that you might find helpful.

Visual preference	Auditory preference	Kinaesthetic preference
What can you see?	What does it sound like?	How do you feel?
What is around you?	What can you hear?	What can you feel?
Who else is there?	What are you saying to yourself?	What are you doing?
Is the image sharp and clear with bright colours?	What are others saying to you?	Are there tastes or smells?

Where are you in respect of your compelling outcome? Does it seem near or far away?

Now step off the timeline and write down your thoughts in the space below.

_____

_____

_____

_____

When you've done that, step back on the line where you were, today. Look towards the end that represents old age and stand on a place along the line where you feel you will have achieved the work goal you have in mind. On that spot, close your eyes again and imagine you have achieved it. Imagine you are at that time in the future as if it is now. Experience the feelings now of having achieved it. This is called 'associating into it'.

Visual preference	Auditory preference	Kinaesthetic preference
What can you see?	What does it sound like?	How do you feel?
What is around you?	What can you hear?	What can you feel?
Who else is there?	What are you saying to yourself?	What are you doing?
Is the image sharp and clear with bright colours?	What are others saying to you?	Are there tastes or smells?

Now we are going to 'anchor' this experience of having achieved the compelling outcome. Associate into this place by making the images brighter and the sounds clearer. Imagine you are in a film of your life, so turn up the colour and volume. When the feeling is really strong and present, squeeze your earlobe to anchor it. By doing this we have an action associated with the feelings. As the feelings fade, remove the anchor and, as they come back, replace the anchor so you are only squeezing your earlobe when you are completely associated and in the zone. The more you imagine yourself achieving what you want, the more your unconscious mind believes it already has and the more likely you will be to get it.

Take a break now. We call this 'breaking state'. Come off the line and write down what you experienced.

Now you know what it feels like to have achieved your work goal. Go back to the point you were just on when you had achieved it and look back towards the point that represented today. We are going to break this journey down into little steps. From this point, can you see what you need to do to get to where you are now? Write the steps down here. This is our strategy and we will cover it more fully in the final chapter, Chapter 14.

1 _____

_____

_____

2 _____

_____

_____

3 _____

_____

_____

4 _____

_____

_____

5 _____

_____

_____

6 _____

_____

_____

7 _____

_____

_____

8 _____

_____

_____

9 _____

_____

_____

10 _____

_____

_____

As you go through the steps there may be areas you need to learn more about. There may also be resources you need, so identify them now. Break down your action plan and take one step at a time towards your goal.

Remember: you have all the resources you require, including the ability to learn new skills and overcome limiting beliefs. Once you have identified what you need, think about where and when you have that resource or from whom you can learn it.

→ I need

_____

_____

→ I have it already when I

_____

_____

→ I can learn this from

_____

_____

As you take each step, use your anchor to remind you what it feels like to already have your outcome; this will keep you focused and motivated.

Use the logical levels, working down from the outcome at the top of the pyramid to consider what changes you need to make at each level, in order to align yourself towards your outcome. You will need to:

- check that your identity fits with your compelling outcome
- check that your beliefs and values support it
- check that your skills are in place or at least resourced, if you need to learn new skills
- check that your behaviour is in line with what you want to achieve, changing it if you need to
- check what impact what you do will have on your environment, and change that too if necessary

Consider and work through each level in turn to align yourself. Here is the pyramid again to refresh your memory.

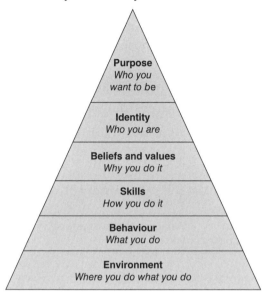

Purpose
Who you want to be

Identity
Who you are

Beliefs and values
Why you do it

Skills
How you do it

Behaviour
What you do

Environment
Where you do what you do

Exercise 45

Sport/activity

Let's work now on something you want in your life outside work. Do you play a sport? Do you do any activities? Choose one that you'd like to work on today and write it in here:

_____

_____

Now decide on a positively worded, specific and well-formed outcome that you can control for your sport or activity and write it in the box below. Start with 'I want to be...'

Remind yourself of your timeline and which end represents your childhood and which is your old age. Stand on the place that represents today for you. As you stand there, think about the sport or activity you want to work on today. Imagine yourself doing the activity and focus on it. Here are those questions again.

Visual preference	Auditory preference	Kinaesthetic preference
What can you see?	What does it sound like?	How do you feel?
What is around you?	What can you hear?	Think about the different parts of your body. How do they feel?
Who else is there?	What are you saying to yourself?	What are you doing?
Is the image sharp and clear with bright colours?	What are others saying to you?	Do you feel hot or cold?
		Are there tastes or smells?

Where are you in respect of your compelling outcome? Does it seem near or far away?

Now step off the timeline and write down your thoughts in the space below.

_____

_____

_____

_____

_____

When you've done that, step back on the line where you were, today. Look towards the end that represents old age and stand on a place along the line where you feel you will have achieved the sports or activity goal you have in mind. On that spot, close your eyes again and imagine you have achieved it. Imagine you are at that time in the future as if it is now. Experience the feelings now of having achieved it.

Visual preference	Auditory preference	Kinaesthetic preference
What can you see?	What does it sound like?	How do you feel?
What is around you?	What can you hear?	Think about the different parts of your body. How do they feel?
Who else is there?	What are you saying to yourself?	What are you doing?
Is the image sharp and clear with bright colours?	What are others saying to you?	Do you feel hot or cold?
		Are there tastes or smells?

Associate into this by making the images brighter and the sounds clearer. Imagine you are in a film of your life, so turn up the colour and volume. When the feeling is really strong and present, choose a different anchoring action from the one you made in the previous exercise – such as a thumbs-up sign – but anchor it as you did before. By doing this we have a different action associated with the feelings. As the feelings fade, remove the anchor and as they come back, replace the anchor so you are only doing it when you are completely associated and in the zone. The more you imagine yourself achieving what you want, the more your unconscious mind believes it already has and the more likely you will be to get it.

Break state, come off the line and write down what you experienced.

_____

_____

_____

_____

Now you know what it feels like to have achieved your sports or activity goal. Go back to the point on the timeline when you had achieved it and look back towards the point that represented today. We are going to break this journey down into little steps. From this point, can you see what you need to do to get to where you are now? Write the steps down in the space on the next page.

1 _____

_____

2 _____

_____

3 _____

_____

4 _____

_____

5 _____

_____

6 _____

_____

7 _____

_____

8 _____

_____

9 _____

_____

10 _____

_____

As you go through the steps there may be areas you need to learn more about or practise. There may also be resources you need, so identify them now. Break down your action plan and take one step at a time towards your goal.

Remember: you have all the resources you require, including the ability to learn new skills and overcome limiting beliefs. Once you have identified what you need, think about where and when you have that resource or from whom you can learn it.

→ I need

_____

_____

→ I have it already when I

_____

_____

→ I can learn this from

_____

_____

As you take each step, use your anchor to remind you what it feels like to already have your outcome; this will keep you focused and motivated.

Use the logical levels, working down from the outcome at the top of the pyramid to consider what changes you need to make at each level, in order to align yourself towards your outcome. You will need to:

- check that your identity fits with your compelling outcome
- check that your beliefs and values support it
- check that your skills are in place or at least resourced, if you need to learn new skills
- check that your behaviour is in line with what you want to achieve, changing it if you need to
- check what impact what you do will have on your environment, and change that too if necessary

Consider and work through each level in turn to align yourself.

Here is the pyramid again to refresh your memory.

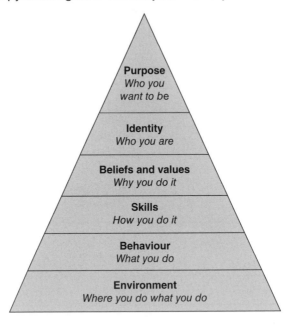

Exercise 46

Your relationship

Let's repeat the exercise for your relationship goal. Whether you are currently in a relationship or not, you will have some ideas about what you want from your personal life in order for it to be more fulfilling. What do you want from your relationship? Write your compelling outcome below. Remember: it has to be something you want for yourself as you cannot control other people. The goal should also be ecological, in that it does not have adverse repercussions for others.

_____

_____

_____

_____

_____

_____

Go back to your timeline and stand on the point that represents today again. This time, focus on your relationship. Here are the key questions again.

Visual preference	Auditory preference	Kinaesthetic preference
What can you see?	What can you hear?	How do you feel?
Where are you?	What are you saying to your partner?	What can you feel?
Who else is there?	What is your partner saying to you?	What are you doing?
		Do you feel hot or cold, or just right?
		Are there tastes or smells?

Where are you in respect of your compelling outcome? Does it seem near or far away?

Now step off the timeline and write down your thoughts in the box below.

When you've done that, step back on the line where you were, today. Look towards the end that represents old age and stand on a place along the line where you feel you will be happy in your relationship and will have achieved the compelling outcome you identified earlier. Experience the feelings now of having achieved it.

Associate into this experience by making the images brighter and the sounds clearer. Anchor it again (using a different action) when you are completely associated and in the zone. The more you imagine yourself achieving what you want, the more your unconscious mind believes it already has and the more likely you will be to get it.

Break state, come off the line and write down what you experienced.

_____

_____

_____

_____

_____

_____

Now you know what it feels like to have achieved your relationship goal. Go back to the point you were just on when you had achieved it and look back towards the point that represented today. We are going to break this journey down into little steps. From this point, can you see what you need to do to get to where you are now? Write the steps down here.

1 _____

_____

2 _____

_____

3 _____

_____

4 _____

_____

5 _____

_____

6 _____

_____

7 _____

_____

8 _____

_____

9 _____

_____

10 _____

_____

Identify the resources you need for each step and break down your action plan towards your goal.

Once you have identified what you need, think about where and when you have that resource or from whom you can learn it.

→ I need

_____

_____

_____

→ I have it already when I

_____

_____

_____

→ I can learn this from

_____

_____

_____

As you take each step, use your anchor to remind you what it feels like to already have your outcome; this will keep you focused and motivated.

Finally, use the logical levels to consider what changes you need to make at each level to align yourself towards your outcome. You will need to check that your identity fits with your new purpose.

Here is the pyramid again to refresh your memory.

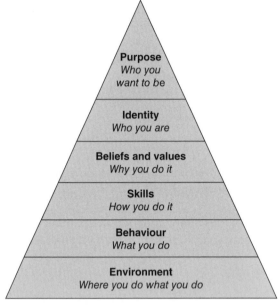

You have now worked through three compelling outcomes using the timeline, anchoring and the logical levels. You can use this process for anything you really want to achieve.

Focus points

- Although SMART goals have their place, we can create compelling outcomes for any situation in life, even for something as seemingly insignificant as an email or text.

- By deciding what we want from each event, we can set about achieving it with passion and belief.

- We cannot control others or their response to what we do, but we can ensure that whatever we do has a positive intention and is ecological, in that it doesn't have an adverse repercussion on someone else.

- By using a timeline and anchoring the associated feelings and experience of having achieved our compelling outcome, we can use that anchor to motivate us as we set about the strategy that will ensure that we achieve what we want.

Where to next?

Often, the only thing that gets in the way of us achieving what we want is our own limiting beliefs. In the next chapter, we will learn how to spot and overcome them. Limiting beliefs usually include phrases such as 'I can't', 'I mustn't' or 'I shouldn't', and what links them to our compelling outcome – or our vision and purpose – is the words **'and what if I could?'**

7 Overcoming limiting beliefs

When we struggle to achieve our goals, it is usually because we have put some sort of barrier there. It may be a belief we grew up with, something someone once said or did or even fear of what might happen when we do achieve our goals. Every time we say we 'can't' do something, this is a limiting belief. Even if we don't consciously say the word, the feeling is there in our unconscious mind and it affects our behaviour.

By doing the sentence completion exercises and questionnaire in this chapter, you will learn how to spot limiting beliefs, and understand why some words can be 'toxic' and limiting, especially when they are in the form of an annoying 'inner voice'. Limiting beliefs occur daily without us realizing it. Even when we tell ourselves that we will 'try' and do something, we are admitting that it will be difficult, even impossible, to do it.

Here's an example.

Exercise 47

Find something to pick up, like a book or a mug or anything readily to hand. Place it in front of you.

Now imagine it is really heavy and tell yourself to *try* and pick it up.

→ How was it? Did it seem quite heavy?

Now tell yourself to *really try* and pick it up.

→ Did it seem heavier this time?

Now take on the belief that it is only a book or a mug and just tell yourself to pick it up.

→ Did it seem lighter and easier to pick up this time?

→ What have you learned from this?

_____

_____

_____

_____

Each time you find yourself using the word 'try', rewind and tell yourself to just *do it*! Similarly, avoid using it with others because you are passing on to them your limiting belief in their ability to do what you've asked. If you ask a colleague to 'try' and get something done by the end of the day, the chances are they won't succeed because they don't think you expect them to be able to.

The unconscious message in the word 'try' is that you (or they) can't do it.

_____

Let's make a list of the things you think you can't do right now!

 # Exercise 48

Draw up a list of all the things you can't do. Then write alongside each one how you know this. What is your evidence?

I can't...	I know this because...
1	
2	
3	
4	
5	
6	
7	
8	
9	
10	

Sometimes our limiting belief stems from childhood. Perhaps someone told you that you couldn't do a certain thing when you were young, a parent or teacher whom you believed knew better than you. These childhood beliefs can taunt us through life and become part of our identity. You might say, for example, 'I am someone who can't do maths' or 'I am someone who can't write well'. Now is your opportunity to revisit these old beliefs about your ability and decide whether they are serving you well today. Maybe now is the time to change them for good?

Exercise 49

Take one of your limiting beliefs from the last exercise. Which is it? Write it down here.

I can't _____

_____

_____

Now write down how you know this, again from the previous exercise.

_____

_____

_____

On the same imaginary timeline that we used in Chapter 6, go and stand on the place on the line when you first had that belief. When was it?

Associate into it by focusing on how you came to have the belief. Maybe there was an incident or a discussion, a decision that was made or some event that crystallized the issue for you in some way? Once you've identified it, break state and write it down in the box below.

Now go back on the timeline to that same spot and imagine you can hover above the line, above yourself, in fact, and look down on yourself standing there on the imaginary line at that precise moment when the belief was born in you. In this moment, you are disassociating. What do you see? What do you hear? What do you feel?

See	Hear	Feel

How can you revisit that limiting belief now as an adult?

Go back to your timeline and stand on today's point, looking back to yourself as a child. What do you now feel about that belief? Do you really want it or can you move forward and leave it where it belongs? Write down in the box below how your feelings have changed and what your new belief is about that thing.

When we believe something to be true, we tend to act as if it were true. We adapt accordingly at the behaviour and environment level, avoiding situations that will challenge the belief. For example, if you have a belief that you are scared of being underground, you will avoid the tube and if you have a belief that dogs will bite you, you will avoid being near them. By challenging the belief and doing the thing you believe you are scared of, you would replace the limiting belief with a resourceful belief which may be that most dogs don't bite and taking the tube is probably safer than the bus or walking.

Some limiting beliefs operate at the skills level: believing you can't do something because you don't yet know how to do it, such as speaking a foreign language, getting a hole in one at golf or running a marathon. When you find yourself saying, 'I can't...' or 'I don't know how to...', add 'yet' to the end of the sentence because you could learn how to do what you want to do. It's one thing not to be able to do something you don't actually want to do but quite another not to be able to do something you want to do and could learn.

- **Which of the limiting beliefs you cited at Exercise 49 are things you could learn how to do?**

_____

_____

- **Which of these would you like to learn how to do?**

_____

_____

- **Think about how you could do this, who you could learn from and when.**

_____

_____

- **When you learn how to do this, what will it enable you to do that you haven't been able to do before?**

_____

_____

_____

_____

Do you notice the word 'when' rather than 'if' in the last bullet point? There is a reason for this. The word 'if' presupposes that it may not happen or it may not be successful, whereas the word 'when' assumes it will be. 'If' represents another limiting belief, rather like the word 'try' mentioned earlier.

Parents use these words all the time with children, suggesting that *if* they go to bed nicely they will get a treat or *if* they eat up all their vegetables they will get a pudding. This is often followed up by an instruction to *try* and eat them up or to *try* and hurry up. Imagine if we removed those limiting beliefs and instead said, 'When you have gone to bed nicely you will get a treat', or 'When you have eaten up your vegetables you will get a pudding'. Here we are now suggesting that we believe they will do it.

It is the same in a work context. We might say, 'If you could just try and tidy up that chart', or 'If you could try and get that report written by close of play today', whereas we would have greater chance of a successful result by replacing those words with 'When you've tidied up that chart we can send it off to the client', or 'When you've finished that report by the end of the day, put it on my desk'.

Exercise 50

According to NLP, there are three main reasons why we limit ourselves. Which do you recognize in yourself?

A   Hopelessness – I don't believe it is possible.

B   Helplessness – It is possible but not for me.

C   Worthlessness – I don't deserve it.

Go back to Exercise 48 and mark alongside each of your limiting beliefs which of these three apply.

→ How many 'a' answers?   ☐

→ How many 'b' answers?   ☐

→ How many 'c' answers?   ☐

What do you learn from this?

_____

_____

_____

_____

_____

_____

Limiting beliefs are your choice. It is up to you whether you want to keep them or get rid of them. The last few exercises were designed to help you do this.

Go back to the beliefs you wrote A alongside, saying you did not believe it was possible. Do you still believe that it is not possible or can you use the NLP timeline to challenge what you wrote? Have you ever believed it was possible?

If you don't believe that a thing is possible, has anyone ever done it? If one person can do it, then you can learn to do it yourself by modelling them (which we will cover in Chapter 10).

Go back to the Bs, where you say it is possible but not for you. How is it not possible for you? What would have to be true for it to be possible for you? Can you now consider that it is possible for you? In that case do you want it and if you do, how could you make it possible? Maybe you could look at compelling outcomes in the last chapter again.

The Cs are about not being worthy. You believe that you are not worthy to get that promotion or that thing you want. Again, what would have to be true for you to be worthy? How can you bring about that change? Have you ever been worthy? When was that? Go back to the timeline to find out when, associate into the feeling, anchor it and bring that into your life today.

Exercise 51

Think of a belief you want to change. Which belief do you have that holds you back in your life from achieving what you want? Write it below.

1 _____

→ How is this belief a problem for you?

2 _____

→ When does it affect you most?

3 _____

→ What is it that you are currently not able to do because of this problem?

4 _____

→ How much do you want to overcome it on a scale of 1–10, with 1 being not at all and 10 being more than anything in the world?

5 _____

→ What is the positive benefit of having this problem?

6 _____

→ When this problem disappears what will you have to do that you are currently not doing?

7 _____

→ Do you believe that this problem is within your control?

8 _____

→ What is stopping you from changing this belief?

9 _____

→ How will you know when the problem is gone?

10 _____

→ Imagine now that the problem is gone. The belief you had is in the past. What do you see, hear and feel that makes you sure it has well and truly gone?

11 _____

_____

_____

Beliefs act as a self-fulfilling prophecy. If we believe something negative about ourselves we tend to be on the lookout for it. We notice it whenever it happens. For example, say you had a limiting belief about your ability to give a good presentation. As you stumble on your words, your inner voice says, 'There you go, you're useless'. Is it any wonder, then, that with that limiting belief in your mind you stumble again and get flustered? If instead you replaced this limiting belief with one that said, 'I am good at presentations', even if from time to time you stumble occasionally over the odd word, you and your audience wouldn't notice because the resourceful belief that you are good will dominate.

_____

Here's an example of how the thought process works.

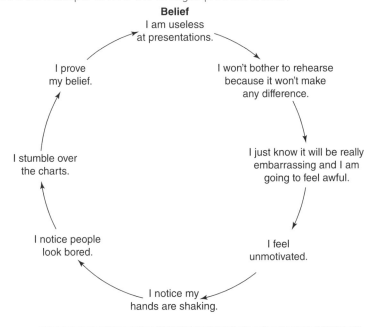

Some great ways to get rid of a limiting belief include the swish and the postcard.

Exercise 52

The swish

First, think about something you want to change. Make it something different from the belief you chose for the previous exercise; you are using this workbook to work on your different limiting beliefs. Write what you want to change in the box below.

Now think about what the trigger is that sets the negative belief into motion. What is the starting point? In the example above it might have been your boss telling you that you have to give a presentation, or simply seeing the word 'presentation' in your diary. Whatever your trigger for the negative feeling, get a picture in your head of that trigger factor. We call this the **cue picture**.

Give it colour, sound or clarity so you could describe it to someone.

Break state and draw or describe it in the box below. You can use 'stick men' and draw bubbles to write in what each person is saying.

Now decide what you want instead. Use the box below to write in or draw your desired image.

You now have two images. Imagine you are looking at a TV and in the main screen is the cue image (the one you currently get as an unresourceful response to the trigger). In your mind, place the preferred image very small in the corner of that.

It should look something like this, with the thumbs-down image being the negative belief and the thumbs-up the replacement, resourceful belief.

Now make the sound 'swish', accompanying this with a hand movement as if swatting a fly, and make the small image replace the big image, like this.

You will probably find it takes a few practices to be able to do the swish quickly and successfully, but when you can, you can use it for all the images you get in your head that you'd rather replace.

Exercise 53

The postcard

This is another exercise designed to get rid of limiting beliefs. Choose another one of yours and have a go at this.

Imagine that your limiting belief has left you. It has gone away somewhere and you don't want it to return … ever! Write your belief a postcard, telling it what you are able to do without it there to stop you. Tell it what a great time you are having and what you are now doing without it.

So far in this chapter, we have used a number of different exercises to tackle limiting beliefs. Some will resonate with you and others won't. That's fine. This last exercise focuses on the NLP belief that every limiting belief has a positive intention. It is there for a reason and if we can understand its positive intention then we can find a way to meet the positive intention without the limiting behaviour.

 ## Exercise 54

Perceptual positioning

Place three chairs in a triangle and choose one to be first position (this is you), another to be second position (this is your limiting belief) and the third to be third position (this is the impartial or disassociated observer).

- Sitting in first position, tell second position – the limiting belief – why you don't want it.

- When you've said what you have to say, go and sit in second position and imagine that you are the limiting belief and you have a positive intention for first position. There is a good reason why you are trying to stop first position from doing something or being something. Be curious about what it could be as you sit in the second position chair.

- Now return to first position and respond to this. You can switch chairs a few times until you really understand the positive intention and have explored it and responded to it.

- When you have finished, go and sit in third position and comment on what you have observed. How would you suggest, as an impartial observer, that first position could find a way to meet the positive intention without the behaviour? Write your answer in the space below.

_____

_____

_____

_____

_____

Focus points

- Your limiting beliefs are beliefs that interfere and hold you back from what you want in life. They limit your ability to achieve your purpose and your compelling outcomes, both in everyday life and for your life as a whole.

- Words like 'try' and 'if' can be limiting, both when externally said and internally thought. By removing the word 'try' and just doing it, and by replacing the word 'if' with 'when', we offer ourselves and others, possibilities and choices that are resourceful.

- As well as the timeline, we used techniques such as the swish, postcard to a limiting belief and perceptual positioning, all of which allow you to challenge limiting beliefs and entertain the possibility of living without them in the future.

 # Where to next?

The next chapter is about confidence. In the context of NLP, confidence is about being resourceful, operating at your most excellent and being the best you can be. It is about being aware of your strengths and skills, learning from feedback and knowing how to create rapport, all the things you have learned from this workbook so far. Being free of limiting beliefs is an essential ingredient in the area of confidence and the next chapter shows you how to make these beliefs conscious, challenging them whenever they sneak into your unconscious, so that you will find confidence.

Feeling confident and resourceful

8

Feeling confident and resourceful enables us to achieve what we want in life, whether that is through a calm state or an energetic and motivated state. It is useful to be able to get into our desired state whenever we need to. This chapter offers exercises to do just that, and explains how states can be achieved through reframing and anchoring.

How confident do you feel right now? Give yourself a score of between 1 and 10 and write it in the box below. ☐

Now think about how you decided on your score.

- Was it based on how happy or sad you are feeling?
- Was it based on your work status, whether you are in a job you enjoy, or was it based on your state of health, how well you're feeling?
- Perhaps it was more about how many friends you have and how you feel when you're with them, or more about your relationship with your partner or with your boss or your children.
- Did it depend on how well you're doing in your sport or what you do in your free time?
- Was it a function of what you see when you look in the mirror?

 ## Exercise 55

Write down in the space below the factors that contribute to your state of confidence. It doesn't matter if there aren't 10; just write down the factors that make a difference for you personally when it comes to confidence. Write them in any order you like, as you think of them. The factors can be whatever you choose, so feel free to depart from the ideas offered above. Your situation and life circumstances will be unique to you and that's what matters.

122

1 _____

_____

2 _____

_____

3 _____

_____

4 _____

_____

5 _____

_____

6 _____

_____

7 _____

_____

8 _____

_____

9 _____

_____

10 _____

_____

Now go back through your list and think about how important each one is to that confidence score you wrote in the box above. Write a 1 alongside the factor that is most important, 2 alongside the next most important, 3 alongside the next one and so on.

→ Associating and anchoring

In Chapter 2 we explored whether you are visual, auditory or kinaesthetic, and you may find that this is reflected in your order of priority. Someone who is visual will be likely to feel most confident when they look good or when they are happy with the appearance of their home, or perhaps a project they are working on. An auditory-preferenced person will feel confident when they speak well, sing or make good music. A kinaesthetic person will prioritize their feelings and actions, so they will be most confident when they play a good game of tennis or hit a good drive in golf.

As you think about feeling confident, I want you to 'associate' into it. This means really imagining yourself at your most confident. Think of a time when your score would be a resounding 10! It might be on a specific occasion or at a special time in your life. Close your eyes if you need to and imagine it's happening right now.

You might find it helpful to change position, maybe stand up and get your whole body into that state of feeling confident.

Exercise 56

Stand up and give yourself a little shake. This is called 'breaking state' and it helps to begin an exercise like this because, since you've just been reading this book you will be in your reading state. We need a different state to do the exercise because we need to engage whole body association – not just your eyes and brain but your body from the top of your head down to your toes.

Think about the time when you felt really confident and resourceful. Feeling resourceful means knowing instinctively what to do and how to do it well. As you think of that time, play it through in your mind as if it is a film at the cinema. Put yourself in the lead role with whoever is important to that feeling of confidence around you, as fellow actors on the screen.

Give the film colour and sound, and music, too, if that helps. Look around the set to what else is in the scene and who is saying or doing what. What colours can you see, what can you hear, what is happening?

- **As you focus on this, be aware of how you are holding your head. Adjust it until it feels aligned to your image of confidence. Be aware of your facial muscles, give them a stretch by smiling widely, and** then adjust them so they also align with your feelings about this confident time. What about your chin? Is it down or up? Adjust it so that it feels right to you.

- Your shoulders, are they tense? Relax them and give them a little shrug before setting them until they feel comfortable.
- What about your arms and your hands? How are they? Hold them in a way that reflects the confident pose.
- How is your core? Is it aligned? Are your hips and back comfortable? Breathe in and out a few times and as you exhale let your stomach expand with air. Then pull in your stomach, pushing the air out as you do so.
- Now think about your legs and feet. How are you standing? Let these confident feelings go into your legs and stand as you do when you feel really confident.

Once your whole body is living this confident feeling, be aware of how it feels and, if you can, take a look in the mirror, take a photo of yourself and keep it somewhere where you can easily take it out to remind yourself how you are when you are confident.

Getting into the confident body position will usually get you the confident feeling and the rest you can do with your mind, recalling the confident time and situation.

Now break state for a few minutes, as that was probably quite exhausting! Give yourself a shake and change your body position by walking about a bit.

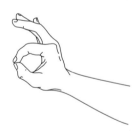

We need an **anchor** now. This is a physical action that will usually remind us of this confident time. It could be linked to how you stood when you were confident. Some people squeeze their ear lobe as an anchor because it's quick and easy to do anywhere without drawing attention to yourself. Others put their thumb and first finger together like the 'OK' sign. Decide what will be your anchor and practise it a few times. You will already have anchors from earlier exercises, so make sure you have a different one for each feeling you want to anchor.

Now remind yourself of that confident time, relive the moment, run the film you created before with the same people, visuals, speech, actions, music and so on. Get into the body position you had before, checking in with each part of your body so that you are completely associated and aligned with that confident feeling. When you have it 100 per cent, anchor it by using the physical action you have decided upon. Keep the anchor in place while the feelings of confidence are strong but, as they fade, take the anchor away so it is only in place when the feelings are there. You want to anchor the strong feelings, not the fading feelings.

Break state; you know how to do that now! Then repeat the previous paragraph a few times until you're sure that when you apply your anchor at a time you need that confident feeling, the anchor will trigger it for you reliably.

You now have an anchor for confidence. You can do what we call 'stacking anchors' to make that anchor even more effective by using the anchor whenever you experience a strong, confident or resourceful feeling in any part of your life. This will make the anchor really strong.

Now just check in with your confident score again. What is it now out of 10? I hope it's a lot higher than before! ❏

If you want it to go higher still, look at some of the other factors you listed at the beginning of the chapter and use the anchoring exercise again, this time with different confident situations based on other factors. Sometimes we can be mistaken about the times we are confident because we perhaps notice more the times when we don't feel confident. So go back to your list and focus on times when you are confident.

→ Reframing

Another way we can build our confidence is by reframing. This means that we take a situation and imagine that it is a picture in an ugly frame (the frame represents the way we are viewing the situation) and we are going to now take that same picture and put it in a much more attractive frame, metaphorically.

So how do we do this?

Behind every behaviour is a positive intent. This is a basic NLP presupposition and it means that our brain does everything with a purpose, even if it is subconscious, which it usually is. Sometimes things don't work out as we'd wish and, in those circumstances, we can put it in a positive frame, give it a positive spin and use it to rebuild that lost confidence.

Reframing also gives us the opportunity to be quite creative by imagining how we can possibly put a situation in a positive light or find the positive intention in a situation that might otherwise seem negative.

Example

Recently my daughter had quite a serious car accident, when she suffered concussion and whiplash. This was certainly a negative experience, but she reframed it by recognizing the importance of driving slowly (she would have been killed if she had driven any faster), wearing a seat belt and having a car with good safety features as opposed to choosing a car on appearance alone. From now on she will drive differently and be more cautious. She has learned from a bad experience and that was its positive intention.

Another way of doing this is through humour. When we laugh at something and find the humorous side of what happened or what was said, this gives it a positive twist. Here's an example of something that happened to my husband and me some years ago.

Example

We were invited to a 'Hello party', which was meant to be an opportunity for some people newly arrived in the area to get to know their neighbours. However, I assumed (being a *Hello* magazine reader) that it was a fancy-dress party where we all had to dress up as a celebrity. This was partly because the invitation was in the form of a collage of celebrities. Anyway, we dressed up as rock stars and duly arrived at the party with big grins on our faces, only to be greeted at the door by the other guests in normal party clothes.

At that point we could have put the situation in a negative frame and been hugely embarrassed and possibly gone home to change. We didn't. We all laughed about it and were the life and soul of the party. People still remember the event and chuckle about it.

Framing is about making a choice. Humour is one such choice.

Another choice is to disassociate and see the situation from another perspective, either someone else's point of view specifically or how an impartial observer might see it. This separates our own emotion from the situation and enables us to see it in a different context. Time and memory can do this for us sometimes. We might look back on an embarrassing presentation we gave or a time when we did something foolish. Yet when we look back and put this in the context of our whole life, we can make it diminish in importance and relevance to our life today.

Exercise 57

This exercise is called **the six-step reframe** and shows how to go through the reframing process for yourself.

1 First, think of an experience you want to reframe. This could be something you wish hadn't happened or something you didn't do very well. It could be a job interview, review with your boss, discussion with a teenager or a date. It is important to choose something that was significant to you in that it affects your confidence today.

2 Now imagine you are talking to that part of you that holds feelings of shame or unworthiness and ask it whether it is willing to let these be resolved and whether it will work with you. Sometimes we can hold on to a negative experience because it actually benefits us in some way. Perhaps being bad at something has a positive intention – to give you the excuse not to have to do it ever again. In this exercise, therefore, we need to establish that you want to change, so this is why we have to acknowledge the feelings and ask ourselves whether we are ready to re-examine them and move on.

3 Ask the part that holds this negative experience what its positive intention was. How was it trying to help you or support you? What positive learning did it have for you?

4 Now ask it to think of three other ways in which it could have given you the positive learning or experience.

5 Evaluate each one and decide which of them would be the best way for you to take that positive learning.

6 Check that this is feasible.

Many people are fearful of giving presentations at work and simply tell their boss that they 'can't'. Maybe they had a bad experience or perhaps they grew up being told they were shy? This may hold them back in their career but that may be what they want subconsciously: to stay safe and unchallenged. However, if it isn't then we can explore the positive intention.

Even famous actors get stage fright and they claim that this sharpens them and improves their performance. We can get nervous even when something quite exciting is about to happen, like a wedding. If we reframe the nervousness as being for our benefit, to improve our performance, and then go on stage and give that presentation from a different frame, we can turn the nerves into a positive and give a great show.

By reframing our disappointments as learning experiences and by taking their positive intention for us, we can build our confidence and do more than we ever dreamed possible.

I want to work on my confidence in the area of

_____

_____

Note down what you do.

_____

_____

Results

```




```

Focus points

- Because what we focus on is what we get, one way to build confidence is to focus on – or associate into – a time when we felt confident; if you had been asked to think about when and where you _didn't_ feel confident, you would have felt that lack of confidence.

- If we notice when we are at our most resourceful, we can anchor that feeling, making sure that each anchor we use is different for each feeling we want to anchor.

- Another way to build our confidence is by reframing a difficult, uncomfortable or challenging situation, so that we see it in a much more positive light.

- We can choose to reframe either through humour or by disassociating and seeing the situation from another perspective. This separates our own emotion from the event and enables us to see it in a different context.

- If we think of an area in our life where we would like to be more confident, we can put it into practice using our anchor and the six-step reframe process.

 # Where to next?

The next chapter is about tackling fear. An NLP principle states that behind every behaviour is a positive intention, and there is a positive intention behind fear. How can we learn how to harness the positive intention behind our fear and use it as a resource? Even successful athletes and actors feel fear before they perform, but claim that this gets the adrenalin going and improves their performance. Sometimes fear can be completely debilitating and more akin to a phobia when we can barely function, so the next chapter also offers a 'fast phobia cure' that will get rid of these fears so that you no longer need to avoid situations that cause you to panic.

Overcoming fear

Feelings of fear can be somewhere along a continuum, from being debilitating at one extreme to simply inconvenient at the other. Overcoming fear empowers us to improve our performance, whether that is at work, socially or in our sporting activities. NLP offers specific techniques such as the swish and the fast phobia cure, described in this chapter, which will enable you to tackle your fear.

→ The positive intention of fear

A fundamental principle of NLP is that behind every interaction is a positive intention, so the fear we experience is in some way performing a positive role. So what could it be?

In many circumstances, fear keeps us safe from physical danger. If you were standing at the top of a cliff with a steep drop down to a stony beach, fear of coming to harm, possibly dying even, prevents you from going near the edge. Similarly, fear ensures that we run away from a fierce animal that's charging at us or a fire that threatens to engulf us. By examining what happens in those situations, we can come closer to understanding how fear works to protect us.

Many people experience fear around small animals such as flying insects, spiders, wasps and mice. They know that, with the exception perhaps of a few poisonous specimens in certain countries, most will not actually cause them any physical or emotional danger, so what is that all about?

Some types of danger are not immediate physical danger but may be perceived emotional danger, such as the danger of commitment for fear of being emotionally hurt. For some, it might be the fear of embarrassment, being humiliated or generally moving out of the comfort zone by expressing a controversial opinion, speaking in a conference or important meeting or making the best man's speech at a wedding.

Before we look at the structure of fear and deconstruct it, do the following word association exercise.

Exercise 58

Complete the following phrases.

→ When I think of fear I get...

a picture or image of a

_____

the sound or noise of a

_____

the sense or feeling of a

_____

the smell of

_____

the taste of

_____

→ **and I want to**

_____

What you did there was 'associate' into the fear. You recalled it, experiencing today the feelings you get whenever you encounter your fear, and you acknowledged how you respond.

Disassociating from fear

In order to get rid of the fear it will be necessary to 'disassociate' from it and view it as if it is distant and not a part of you but a separate 'thing' that you can pick up and examine at a conscious level then choose to set aside in exchange for a more resourceful response. By resourceful we mean a response that enables you to do what you want and not let the fear limit you.

We want you to get rid of it – yes – get rid of it! You learned this fear very quickly and we can get rid of it quickly, too. By the end of this chapter, your fear will have gone.

First we see, hear or feel something – which we can call a 'trigger' – that we have a belief about. This belief is that this thing is dangerous and could harm us. The belief may stem from our childhood or from films or TV, where we have learned that this animal or fire could be dangerous. Maybe we have seen how others have reacted and we believe we should do the same. This is how beliefs are passed on. If a parent was scared of dogs, was bitten by one or refused to have one as a pet because of a belief that they were fierce, then you as a child will have taken on this belief too. The trigger switches on the belief; the response happens unconsciously as we step back from the cliff or cross to the other side of the road if we see a big dog.

We all have fears and anxieties because that is how we protect ourselves from danger. However, these fears also limit us because not everything we fear is actually dangerous. For example, fear of flying may have resulted from a bad experience, turbulence perhaps or seeing an aeroplane crash, but it is not a realistic fear. Most aeroplanes don't crash and nowadays not being able to fly because of fear will prevent us taking holidays abroad, making business trips and might even stop us visiting friends and family. So how do we decide which fears are realistic and which are not?

Knowledge is important. If you live in an area where there are deadly snakes, dangerous animals or violent gangs, then your fear of them is justified. However, you can offset this fear by learning how to recognize dangerous creatures or situations, which streets to avoid and how to respond to danger. This puts you in control and allows you to modify your belief down to a manageable level.

Fear of things that you might encounter on a daily basis, such as birds, spiders or flying insects, can be tackled in the same way. Know which could harm you and learn to recognize them. This way your belief changes from 'I am frightened of all birds' to 'I am frightened of this bird or that bird'.

Other types of fear that we meet regularly include fear of failure, fear of change and fear of rejection. These, too, are based on past experiences, beliefs passed down from parents and teachers and, like all beliefs, they can be changed to allow us more choices in life.

You may be anxious about specific things such as giving presentations or speeches, speaking up in a meeting, or making that sales call. Do you avoid certain activities or challenges at work so you don't have to tackle that fear?

The way we can change the response or the behaviour is to change the belief. We can change our belief through increased knowledge and new experiences.

First, let's look at some of the fears you may have.

Exercise 59

This exercise asks you to list all the fears you have, no matter how trivial they might be or how severe and debilitating. First, list them all and then go back through them and give them a score of 1–10, depending on the level of fear they represent and then score them again according to how inconvenient the fear is for you.

I am scared of....	1–10 score for how scared	1–10 score for how inconvenient
1		
2		
3		
4		
5		
6		
7		
8		
9		
10		

Which fears did you rate as the most inconvenient? Let's tackle those first. Take one of them to use for the next exercise.

I am going to tackle my fear of

_____

_____

_____

Exercise 60

Using the timeline again, imagine your line on the floor and stand on the point when you did not have this fear. When was that? You have not always had this fear. Babies are not born with fear; they acquire fears through experiences and what parents and carers tell them. There was a time in your life when you did not have this fear. Associate into this fear-free time by putting yourself there mentally and experiencing without fear the thing you have chosen to work on.

What is it like not to have this fear? Write down your thoughts in the box below. How do you feel being free of this fear?

Notice how you are standing and be aware of your body posture. Do you look and feel confident and safe?

→ How are you interacting with the thing that you fear now?

→ What is your belief about this thing?

I believe

_____

_____

_____

Now take a step forward to the point when you first experienced the fear. When was that and can you recall what the trigger was? What prompted you to have the fear for the first time?

What happened was

_____

_____

_____

_____

_____

_____

_____

Can you take yourself above the line mentally and look down at yourself experiencing what you experienced? You need to disassociate from the fear. What do you see? What can you hear? What do you feel? What belief do you think has replaced the belief you had earlier?

_____

_____

_____

_____

_____

Still looking down on yourself, can you find a way to tell the 'you' who is standing on the line that this new belief has a positive intention? What was the positive intention of the fear you experienced then? Write it down.

The positive intention of the fear was to

_____

_____

_____

_____

Now go and stand on the point that represents today and consider other ways you can satisfy the positive intention. What can you do differently? It may help to adopt the body posture you noticed at the point before you had the fear. Taking on the physiology is a way of taking on the belief. What other choices do you have now? Write them down in the box below.

As a result of doing this exercise, is there more you need to find out, more experience you need or training that will enable you to meet the positive intention another way? For example, by being better at doing something we can overcome the fear of doing it and failing, because we have put some expertise in place.

Example

Mel was terrified of going down black runs (the most difficult slopes in skiing). She would stand at the top watching everyone go down with skill and imagine herself having a nasty fall at speed and ending up in hospital. She had a negative belief that she was unable to ski well enough to do black runs despite having skied for many years very competently. She associated the word black with danger and death. Sometimes she had been down black runs without realizing and skied them well. It was knowing that the run was 'black' that was the trigger.

When she used the timeline and looked down on herself, she saw someone making a fuss about nothing. The positive intention of her fear was to keep her in her comfort zone. She realized that she could stay in her comfort zone and ski happily down black runs by getting more practice at 'edging' (digging the edges of the skis into the snow) to gain more control. Her new belief was that she could do it.

Let's look now at a fear that you have scored highly for it being extremely scary. For this one we are going to use the **fast phobia cure**. This allows you to re-experience your phobia or fear without also experiencing the emotional content. Remember that you were safe before you experienced the fear and safe afterwards.

Exercise 61

I am going to tackle my fear of

_____

_____

- Identify which fear you want to get rid of and give it a score of 1–10 to show how fearful you feel about it right now. What is that score? ☐
- Imagine you are sitting in a cinema watching yourself on a movie screen with the images in black and white.
- Now imagine yourself floating up to the projection room and watch yourself watching the screen. So you are now watching yourself, watching you on the screen.
- Run the black and white film of you, from just before you experience the fear, through the experience of the fear itself and to the time afterwards when the fear has gone and you feel safe.
- Freeze the film at the end of the sequence and float down from the projection room and place yourself on the screen at this point at the end of the film where you feel safe.
- Now run the film backwards very quickly, this time in colour, to the beginning before you experienced the fear and are safe again.
- Keep repeating this process of reliving the experience backwards from the end to the beginning in colour, fast. You may need to do this a few times to get the hang of it.
- What is your fear score now? ☐

This score can be lower still if we work on it some more and then go back and repeat the exercise at the end of the chapter.

Perceptual positioning

You know you can change your beliefs and we have worked on this already, using the timeline and the logical levels. Another way is to explore the positive intention some more, using perceptual positioning. You will have done this exercise around limiting beliefs in Chapter 7 and we are using it again for fear.

Exercise 62

- Place three chairs in a triangle and choose one to be first position (this is you), another to be second position (this is your fear) and the third to be third position (the impartial observer). This is called perceptual positioning.

- Sitting in first position, tell second position why you don't want it – the fear. When you've said what you have to say, go and sit in second position and imagine that you are the fear and you have a positive intention for first position. There is a good reason why you are trying to frighten them. Be curious about what it could be as you sit in the second position chair.

- Now return to first position and respond to this. You can switch chairs a few times until you really understand the positive intention and have explored it and responded to it.

- When you have finished, go and sit in third position and comment on what you have observed. How would you suggest, as an impartial observer, that first position could find a way to meet the positive intention without the fear?

Example

Geoff is a writer who has spent the last six years writing a novel. He wants to submit his manuscript at the York Festival but fears rejection. He has always held back from being judged in his life. He does the perceptual positioning exercise and realizes that the fear of rejection has a positive intention. It wants him to stay in his comfort zone and feel safe. He decides that, while this suited him in his youth, he now wants to be judged and is able actually to welcome it. Being judged has been important to him in his writing career and he replaces the fearful feeling with excitement. Geoff goes to the York Festival and is offered a two-book contract. His first book is then published to great acclaim.

Fear and excitement often go hand in hand. Actors and athletes experience both emotions before going on stage or before an event, and they often report that the heightened awareness and fear set their adrenalin going and improve their performance.

The swish

Experimenting with the swish that we used in Chapter 7 is another way to banish fear.

Exercise 63

Imagine the thing you fear as an image on a screen. Make it big and colourful, as horrible as you like!

Now change the picture by changing the image to your favourite colour, make it look cuddly and appealing, give it a smile, make it smaller, do whatever you need to do to make you smile at it. Then place that image in the corner of your screen. You should now have the nasty image in the centre and the nice cuddly one down at the side, like the picture to the left.

Now make the sound 'swish', accompanied by a hand movement as if swatting a fly, and make the small image replace the big image.

You will probably find that it takes a few practices to be able to do it quickly and successfully. When you can, you can use this swish for all sorts of images you get in your head that you'd rather replace.

Submodalities

In NLP terms, submodalities are distinctions of form or structure (rather than content) within our internal represention. For example, within the visual representation, external and mental images will be either coloured or black and white, and still or moving. These are *submodalities* within the visual sense. As with swish, we can use submodalities to change negative images, sounds and feelings to ones we could cope with.

Exercise 64

It is possible – as you found in the previous exercise – to create an image associated with your fear and then change it consciously so that it no longer makes you fearful. Here is a list of ways you can do this. Change one thing at a time to determine which one makes a difference. Before moving on to another submodality, put the thing back to how it was so that you control the changes one at a time.

- **Location** – move the image, sound or feeling further away in your mind so it looks smaller.
- **Volume** – turn down the sound, change it to mono and alter the pitch and tone so it is less threatening.
- **Colour/black and white** – change the colour so it looks appealing.
- **Associated/disassociated** – remove yourself from the image so you are not involved and can be dispassionate about it by removing the emotion from the experience; for example, 'It's just a mouse', 'It's just a black run'.
- **Size** – make it smaller, shallower or less steep.
- **2D or 3D** – make it two-dimensional so it looks flat and less frightening.
- **Brightness** – make the picture dull and boring.
- **Still or moving** – freeze frame so it is still.
- **Shape** – change the shape.

The key to tackling fear and anxiety, phobias and so on is to make the unconscious conscious by becoming aware of the trigger, acknowledging the underlying belief and changing it so that the response also changes.

Remember that you have encountered the thing you were scared of before and you are still alive. By avoiding tackling it head on and getting rid of the fear for good, you will always have that fear in your life. Imagine your fear as a mountain peak.

As you go up the mountain of your fear, your fear increases but if you run back down the mountain you will never reach the peak. From the peak, you can see everything and then you can descend knowing that the fear is over. If you never reach the peak it will always be there and you will never know what it's like to have scaled the fear and succeeded in overcoming it.

Focus points

- Instead of avoiding fearful situations, we now have the resources to make different choices.
- We can use a variety of techniques to challenge our fears and phobias, including a timeline to disassociate, perceptual positioning to check out the underlying positive intention, swishing to sweep the fear away and submodalities to change the negative images, sounds and feelings to ones we can cope with.
- We can use anchors to gain a feeling of calm and confidence, to use whenever we expect to face anything we find fearful.
- Our unconscious receives instructions from our conscious mind to respond to the trigger for fear. This is a negative anchoring strategy – for example, we see a spider, our eyes widen (visual), we scream (auditory) and we run away (kinaesthetic).
- By using the techniques you now know, you can choose a different result consciously, which will not trigger the fear response.

 ## Where to next?

The next chapter goes to the heart of what NLP is all about. The essence of NLP, what makes it unique among therapies, is the modelling process. In order for an NLP practitioner to become a 'master practitioner' they have to model a skill in someone who has it with excellence and be able to pass it on to someone else, so that they have the structure to get the same result. For modelling, you will be using all the skills you already have and acquiring more.

10 Modelling: acquiring skills from others

An NLP principle is that we have all the resources we need to do whatever we want to do. Perhaps we have a resource in one area of our life and need to apply it in another, or perhaps we have observed it in someone else and need to learn it for ourselves. Modelling is the NLP process of acquiring skills by taking on the detailed structure of the desired behaviour that we have observed, including, importantly, the beliefs and values underlying that behaviour.

→ The concept of modelling

Modelling is the process of coding talent. Essentially what we are doing when we model someone is stepping into their shoes and reproducing what they do, to get the results they get. In order to do that we have to understand the thinking, language and behaviour patterns of the other person, which requires us also to take on their beliefs and values. Proof that we have the model of excellence is that we can pass it on to someone else so that they have it and get the same results.

Modelling is a concept unique to NLP. It is based on the principle that not only do we have all the resources we need to do what we want but that one of these resources, the ability to learn to do something that someone else can do with excellence, is one of them. In fact, we have been doing this quite naturally all our lives from the moment we learned to walk and talk as babies. We learn from our failures as well as our successes, and being able to take on the concept of seeing failure as feedback and a learning opportunity is intrinsic to the modelling process.

Example

Think about the first time you did something like threading a needle.

You might wet the end, hold it up to the light, cut the end of the thread off, try a needle with a bigger eye, or hold it against something dark. You try lots of different approaches to get the thread through the needle.

You could also ask someone to help, perhaps asking them to hold the needle for you. You could ask them how they do it.

However, if you asked a seamstress or tailor who threads needles regularly you would get an expert view on how they do it, assuming they do it with excellence. Beware making the assumption!

Ask them what they believe about threading a needle. Is their belief the same as yours? Probably not: they think it is easy because they do it all the time. If you took on that belief, would you also find it easy?

However, what is also quite a powerful concept is that we can also duplicate something we do ourselves with excellence in one area of our life and transfer it to where we need it in another area.

What skill do you need to thread a needle? Patience, perhaps? Delicate movement? A still hand? When do you have these? Find where you have the skill. Get the belief that you can do it and test it out.

Let's start by making a list of those things we do 'with excellence'. Think of a thing that you do really well, to the extent that people remark on it, you are known for it among your friends or work colleagues, or your family associates you with it. What you do with excellence could be something seemingly relatively trivial such as being able to do a cartwheel or whistle, or it could be something apparently more significant such as running a successful business. Do not be deceived, though. There are significant skills in seemingly minor actions. The more often we do something, the easier it becomes for us, so that we don't even have to think about it. We tend to give more weight to those skills that take us time to acquire or are difficult at first. The map is not the territory, and what may be easy for us may be impossible for someone else.

Exercise 65

What do you do with excellence? Make a list below.

At work	In my sport	At home	Socially

Look at your lists and think about what transferable skills you have. You need to be thinking along the lines of, 'To do this thing well I use this skill, which if I transferred to that area of my life would mean I could do X'. For example:

a I can reach up and put a glass on to a high shelf.

b I believe that I can stretch up high with my arm.

c I could use this skill to improve my tennis serve.

a I can notice when something is out of place on my desk.

b I believe that I notice things that are wrong or different.

c I could use this skill in my proofreading.

a I am always on time to pick up my son from school.

b I believe I have a good sense of time.

c I could use this in my cooking.

Exercise 66

Your turn! Following on from the example above, can you write out some sentences that demonstrate the skill, what it means and how you could transfer it?

→ I can _____

_____

I believe I

_____

_____

This means I can

_____

_____

_____

→ I can _____

_____

I believe I

_____

_____

This means I can

_____

_____

_____

→ I can _____

_____

I believe I

_____

_____

This means I can

_____

_____

_____

In these examples, the skill you identified is your own and, by extracting the belief and thinking about how you do it, you can replicate it elsewhere.

Now let's move on to modelling someone else. The first step is to decide who to model. It's no good choosing simply the person. What you want is the skill: what they do with excellence. A typical modelling project would be worded like this.

I want to model the way (person's name) does (the specific thing they do).

In Chapter 4 you learned about chunking up and down. When modelling a skill, you need to chunk right down. The specific thing you will be modelling will be just a small part of something, a small chunk. For example, a tennis serve can be broken down into small chunks, such as:

- **posture**
- **the swing of the racquet**
- **throwing the ball up**
- **movement of the body**
- **shifting weight forward**
- **hitting the serve**
- **following it through**

The big chunk is 'the serve' but which small-chunk part of it do you need? Which part do you want to model to improve your own serve?

Now it's your turn to think of a skill you want to model. You can repeat the modelling exercise as often as you like and it could be argued that ongoing modelling will become part of your everyday life once you incorporate it into your personal development. So for your first time, choose something easy to model. Think of something someone close to you does, so you have easy access to them; you will need to watch them closely and ask them questions.

 # Exercise 67

→ I have decided to model the way _____does

_____

There are three phases to modelling.

Phase 1

Watch your model doing the skill you want. Watch their physiology, notice how they are holding themselves. What exactly are they doing? How are they doing it?

You can ask them as they do it how they do it, what follows what? What are the steps involved and what is the thinking or belief about each step?

Discovering the belief

Beliefs influence behaviour. A typical example of this is when someone says 'I can't...' How likely is it that this will become a self-fulfilling prophecy?

Why are they doing it? This often results in answers around people's beliefs and values. They tend to take three forms:

1 Beliefs about what things mean – for example, someone who believes 'life is hard' or 'life's a struggle' will ensure that they make it so (unconsciously).
2 Cause and effect – for example, 'What comes around goes around', or 'If you give out you get back'.
3 Priorities – for example, 'The most important thing in life is to be happy'.

To find out your model's beliefs, use questions like these:

- Why do you do what you do?
- What does that mean to you (here we are chunking up)?
- What would happen if you didn't do that?
- What is it like to do what you do?
- In what way does it empower you?

Once you have their beliefs, try them on for size. Imagine you also have this belief. Now answer these questions:

- **What changes do you notice when you have this belief?**
- **What would you do differently when you have this belief?**
- **What else would you be capable of?**

Now find out why each step is important and what happens if they don't do that step. What do they believe about each step in terms of how it contributes to the result?

Take on the physiology by matching with them at every step. Match from top to toe, facial expression, body posture, breathing and movements.

By combining the belief and the physiology, you will get closer to the model and be trying it on for size to see which parts of the model make a difference to the result. Remember that if what you are doing is not working you need to do something different, so even if at first it seems awkward to be matching your model, remember that you want their result, not your own, so you need to use their model.

What you are getting for yourself is a new strategy. It is like a recipe: you need every part of it to make a good meal; you need the ingredients in the correct quantities and quality; and you need the correct methodology – time in the oven and so on.

Use the VAK when you get the structure, so match their preference by asking questions that will resonate with them, such as 'What do you see when you do...?', 'What do you hear when you do...?' or 'What do you feel when you do...?'

When you write down the steps in your model, note whether they are V, A or K in brackets alongside. You can develop this further by noting whether they are external or internal (i) or (e):

- Vi – an internal image
- Ve – an external image
- Ai – inner voice
- Ae – external voice
- Ki – inner feeling
- Ke – external action

Be curious. You want to be able to replicate what they do, so you need every part of the process: the behaviour, the thinking, the inner dialogue. We want the unconscious to become conscious. Your model of excellence may not have known, until you asked, what exactly he does, what he believes about what he does and how important each step in the process is to achieving the result.

Use the box below to write down what you have learned from Phase 1.

Phase 2

This is where you have a go at making the model for yourself. Copy the process you have discovered from Phase 1 and find out whether you have all the component parts of the structure from your model in order to get the result you want.

How did it go?

_____

_____

_____

_____

_____

If you didn't quite get the result you wanted, go back to the model and work it through with them again, a step at a time, checking that you have the behaviour, thinking and internal dialogue exactly as theirs. As you go through it with them, they may say 'Well, it's more like this', or realize they have missed out something that they do that you haven't done, thought or said.

It can take several attempts sometimes to get the whole of the model and then get the result.

Phase 3

You know if you've got it when you can teach it to someone else and they also get the result, because the process is about coding the excellence and being able to replicate it so that it can be passed on to anyone and get the same result.

You may need to tweak and develop the model several times in order to get it 'spot on'. You may even need to find other models for the same skill. For example, returning to the tennis serve, you may have thought you needed the ball throwing part of the serve and modelled that. If you didn't get the result you wanted, i.e. a brilliant ace, then perhaps you need to model another part of the serve.

When modelling, you can use physical people you know or even people you see on TV such as sportsmen or performers who have the skill you are after. You won't be able to ask them the questions but, by reading interviews with them or their autobiography, you will get an insight into their thinking, beliefs and values. You can buy DVDs of many famous people in action and, by freeze framing, you may be able to break down the action into the chunk size you need to model them.

'If someone can do it, you can do it' is the premise here. Be curious and free of assumptions. Remember that it is their thinking you need to tap into. Your own thinking hasn't got you the results you want, so you need to be curious about theirs and prepared to take on their beliefs as if they were your own. Be curious about what they do differently from you. What is the difference that makes the difference? How do they get the results they get?

We are not looking for the 'whys' but the 'hows'. We need to match our mind and body to that of the model so that we not only get the same results but we get them in such a deep way that we can pass them on. It is the structure of excellence that we need to code and the successful coding of the structure enables the skill to be passed on again and again.

Modelling is fundamental to NLP and is the way we acquire new skills for ourselves, pass skills on to others and transfer skills from one part of our life to another where they are needed. The emphasis is on observation, copying and asking questions to get the belief. It is not dependent on the model of the skill telling you what to do. This approach is increasingly being applied in sports coaching, where NLP modelling techniques rather than verbal instruction are used to coach performance.

Once you have modelled a few skills, using the exercises in this chapter, be bold and take modelling on board as a life skill for every day. Keep it finely tuned by being observant and curious, asking yourself, 'I wonder how he does that?' or 'I wonder what her belief is that she can do this?' You may well find that people are delighted to show you how they do something and, when you explain that you'd like to know what's going on in their head as they do it, they are fascinated and may even ask to model something they've observed in you!

Focus points

- Chunk down to get the exact skill you need – you need to model the skill, not the person.
- Find as many models as you can to get the structure.
- Remove all your assumptions and be curious about their structure.
- Get the underlying beliefs.
- Code the structure.
- Take on the beliefs.
- Try it out and continue coding until you get the same result as your model.
- Pass it on so that someone else can get it too.

 ## Where to next?

In the next chapter, you will learn how to use language to influence others, with integrity. We will reacquaint ourselves with deletions, distortions and generalizations, learn how to get agreement with elegance rather than coercion and how to use clean language, metaphors and embedded questions to overcome conscious resistance.

11 Influencing with integrity

In life we often wish to get others to do what we want, whether we are a parent wanting our child to do his homework, a manager wanting our staff to work at the weekend or a coach wanting the team to up its game. By using NLP linguistic techniques such as metaphors and hypnotic language we can influence people to act in the way we desire with integrity. We can encourage them gently to find a solution for themselves, solve a problem or be led towards our way of thinking with consideration rather than through coercion.

→ The language of persuasion

In this chapter, we shall learn about hypnotic language and how we can use this to influence with integrity. We do not want in our everyday lives to put people into a trance or force them to do something; instead, we want to persuade them through the power of our words. In this way, we can bypass their conscious mind, which wants to do things the way it has always done.

Hypnotic language is so vague that the other person cannot make complete sense of it at a conscious level. It therefore bypasses the conscious, allowing the other person time and space to make of it what they will. There is no detail to get caught up on, disagree with or challenge in any way. It is deliberately vague rather than being not vague in the sense that you can't be bothered to mention the detail. It is not the intention to send people into a trance inadvertently or by using vague language confuse them so much that they drift off! The purpose is rather to allow space for the listener to make their own sense of what is said. You will have the opportunity to write your own hypnotic scripts later in the book, which you can then record and use when you need to influence your own self-talk or encourage others.

In the chapter on rapport you were introduced to generalizations, distortions and deletions that all interfere with clean communication. These are called the meta-model, which is used to enhance communication by getting more detail or 'chunking down'.

- Generalizing means that we take one experience from our past and assume that all other similar experiences will be the same. Phrases like 'All...are...', 'Everyone knows...', 'You ought to...' and 'I should...' are examples of generalizing.
- Distortions come about when we apply our own perceptions and assume it is the same for other people too, as when we say, 'You must be feeling...' or 'They all think I'm useless.'
- Deletions occur when we are vague in our communication and miss out the detail that enables the other person to understand what we are saying.

The **Milton Model** takes these same language patterns but makes them deliberate and by so doing takes the person into a receptive state. When we want to influence others using hypnotic language, we want to 'chunk up'. We use the same generalizations, distortions and deletions but in a different way: we use them to lead the other person to reflect.

→ Words to avoid

Words to avoid – so-called 'toxic' words – are words that interfere with communication. We often use such words without realizing it.

What we say matters. However, how we say it makes the difference between simply communicating what we have to say and inspiring others by engaging with them in all senses. We looked at how VAK and the meta-programmes, discussed in early chapters, are part of this, and we will now go a step further to explore how words, tone of voice and the pace we speak can affect communication.

Try

When we tell ourselves or others to 'try' and do something, there is an implication that we – or they – won't succeed. As a result, our message is counter-productive because it suggests the opposite of what we mean. What we mean is for the person just to 'do it', so why don't we say that instead? We use the word 'try' intending to encourage, cajole or gently persuade the person in the hope they will find it easier than they anticipated. Unfortunately, the word 'try' merely confirms their view that the task is too difficult and the chances are that they will soon give up because, after all, you told them only to 'try', not to actually 'do it'.

For example, instead of saying:

- **'Could you try and get that report done by midday so we can go through it together this afternoon?'**

say:

- 'Please get the report done by midday so we can go through it this afternoon.'

If

By using the word 'if', we are offering the other person a choice. They can do it or not, as they wish. Did you want them to have this choice or do you want them to do what you've asked? If the latter, it would be clearer to communicate using the word 'when' to make it clear that you intend the task to be done.

For example, instead of saying:

- 'If you have time, please could you check the rota for Tuesday?'

say:

- 'When you have time, please check the rota for Tuesday.'

But

The word 'but' negates whatever went before it, so put the words you want to emphasize after it. When giving feedback, people often say what they liked about someone's work or presentation and then take away the compliment by adding 'but...' afterwards. All that will be remembered is the last thing that was said, the negative comment after the 'but'. To give helpful feedback that someone will learn from, put the positive comment after the 'but' or replace it with 'and' instead.

For example, instead of saying:

- 'I really like what you've done with the design, but could you make it a bit bigger please?'

say:

- 'Please could you make the design bigger but I really like what you've done with it.'
- 'I really like what you've done with the design and could you make it a bit bigger please?'

Don't

Our unconscious mind obeys commands and 'don't' is a classic embedded command. For example, if someone said to you, 'Don't think about pink elephants!' what do you do? You immediately conjure up the image of a pink elephant in order to make sense of the instruction. After all, you need to have the image in order to know what it is you have to 'not think about'. You had no idea of pink elephants before, did you? Straight away you have disobeyed the instruction without intending to. We use 'don't' all the time and it's easy to see how – if you remove the 'don't' from the sentence – you are actually making the other person think of what you don't want rather than what you do want.

Focus your instruction on what you do want instead, e.g. Be careful, Bring the report to the meeting, Keep going, Keep this private.

This can also mean that we can use the word 'don't' when we mean 'do'. For example, imagine if a teacher said 'Don't stop talking in class', what might happen? The children would probably stop. If at work we said, 'Don't finish that report today' it probably would be done.

Think about when you use these toxic words and be aware of them in future. Which are the ones you are most prone to using? Write them in the box below and then draw a big cross over them so you know they are not helpful.

→ Mastering hypnotic language

Just as we can embed commands with the word 'don't', we can also embed commands in sentences to act as questions even though they are not strictly questions at all. By using a curious tone of voice, the listener is compelled to find a response, e.g. 'I'm wondering when you'll feel confident enough to give a presentation to the team'. 'You'll feel confident' is the embedded command in this case. Similarly, if you say, 'I'm curious to know what you would like to get from running', the embedded command is 'what you would like to get from running'. In these two phrases are the presuppositions that the person will feel confident and that they do get benefit from running.

Another way of making your language skip the conscious defence mechanism is to assume there is no choice in the question, rather like the example earlier of using the word 'if' and replacing it with 'when'. For example, instead of asking 'Can you spare me a moment to talk?' we say, 'When can you spare me a moment to talk?' we are presupposing that they will.

We can also use the word 'how' to ask a question that presupposes someone can do what we ask. Instead of asking, 'Do you want to answer this question?' you'd ask, 'How would you answer this question?'

Just as we can replace the toxic word 'but' with 'and' to greater effect, we can also use 'and' to link two parts of a sentence to produce hypnotic language. The first part may be factual, e.g. 'I know you have learned this concept and now I'm wondering how you will apply it'. The use of 'how' also presupposes that they will apply it.

The **'yes tag'** is another form of hypnotic language. Examples of a yes tag are 'wouldn't you?', 'can't you?', 'won't it?' and so on. A statement followed by a yes tag prompts the answer 'yes' in all cases. So instead of asking, 'Please close the door', you'd say 'You'll close the door, won't you?'

When we use questions that aren't really questions or embedded commands, they work best when we first get ourselves into an attitude of curiosity and pause slightly before speaking, to mark out that we are curious.

Exercise 68

Now you're going to have a go, aren't you! Here are some sentences that need to be rephrased into hypnotic language.

→ Take your shoes off when you come in the house!

_____

_____

→ I'm looking for volunteers to show this guest around the exhibition.

_____

_____

→ You'll get more out of this report if you read it carefully.

_____

_____

→ Do you want to take a break now?

_____

_____

→ It is important that everyone comes to the meeting so we cover all departments.

_____

_____

→ You gave that client a good outline of the company but you didn't mention our latest deal.

_____

_____

→ Are you going to answer the question?

_____

_____

→ Don't be late tomorrow.

_____

_____

→ Try and explain it again to me.

_____

_____

→ Will you be going to the conference in Vegas next week?

_____

_____

Using clean language

'Clean language' is another form of hypnotic language. It is similar in that you are matching the other person's tone and pace, and their actual vocabulary, but here you are reflecting back their words as questions. Earlier we were adding our own assumptions quite deliberately in order to encourage someone to do something, e.g. 'I'm wondering what time today you will finish that report', which assumes you will finish it today not tomorrow. Clean questions are completely free of assumptions and presuppositions and they are respectful in that we are assuming that they themselves know the answer and we are just the catalyst. They are so vague that they encourage and allow the listener to delve into their unconscious for the answer. We take on an attitude of curiosity and simply reflect back what is said using a questioning tone of voice so that the speaker has to expand further. This is part of rapport, of course, because it is respectful, using the other person's own words and language patterns. We use 'and', plus any of the following:

David Grove's Clean Language questions

Developing questions
- **What kind of X (is that X)?**
- **Is there anything else about X?**
- **Where is X? or (And) whereabouts is X?**
- **Is there a relationship between X and Y?**
- **When X, what happens to Y?**
- **That's X like what?**

Sequence and source
- **Then what happens?**
- **What happens just before X?**
- **Where could X come from?**

Intention questions
- **What would X like to have happen?**
- **What needs to happen for X?**
- **Can X (happen)?**

'X' stands for one or more of the person's own words, quoted as closely as possible.

In a more casual conversational environment or when you are just starting to use clean language and can't remember them, the Lazy Jedi questions are a good first resort. They are:
- **What kind of X (is that X)?**
- **Is there anything else about X?**

By mirroring back the other person's language, the speaker acknowledges that what the other person said has been heard and taken on board. The repetition is also somewhat trance-inducing in the sense that the person doesn't feel challenged and can therefore continue elaborating and exploring their feelings. These clean questions are asked in the present tense in order that the speaker associates into the feeling or the experience as if it were happening today.

Clean questions can be particularly useful when we want to understand client needs or help a colleague overcome a problem at work, encourage someone to move out of their comfort zone or build up their confidence. They are usually, but not always, used with metaphors.

Using metaphors

A metaphor is another way of describing something in terms of something else. We do it all the time; in fact, that's how we make sense of our world, by comparing any new thing to something similar to it. Much of this is subconscious. You might like to think of the unconscious mind as an elephant and the conscious mind as the rider on its back. The unconscious mind is considerably bigger and stronger. Anything that is not literally true is a metaphor. Metaphors can be expressed as words such as:

- **'Mealtimes are a battleground.'**
- **'Life's a breeze at the moment.'**
- **'All the world's a stage.'**
- **'I feel stuck.'**

Or they can be images, pictures or symbols. Many businesses choose logos that are metaphors to express a concept that customers will relate to and aspire to. They choose colours, fonts and pictures that promise happiness, power, health, love and so on. Think of the logos for brands and companies you like. How have they communicated in metaphors? If you were setting up a business what would symbols or colours would you choose to reflect what you offer?

Metaphors tell us a lot about people in a way that a straight question and answer won't achieve. We use the unconscious mind to create our metaphors and they are unique to us. In many ways, they explain who we are and what we believe about our world. A good way to play with metaphor is to ask yourself the following questions:

In what way am I like a tree?_____

How am I like a lion?_____

What part of a car am I most like?_____

How does how I brush my teeth reflect who I am?_____

These just give you an idea of how to become curious about how we live our life in metaphor with one action often being a pattern for how we do other things. I once asked my students how they stacked the dishwasher. One said 'there's only one right way to stack the dishwasher and that's the way

I do it'. Another said 'there are lots of different ways and I do it a different way every time'. Another said 'even though I have a dishwasher, I don't use it because washing up by hand is my way of showing my love for my family'. How do you stack yours, and what does that say about you and how you do other things? How is it a metaphor?

_____

_____

_____

_____

_____

_____

Companies use metaphors, too, in the language they use to communicate both internally and externally. What expressions does your company use that are metaphors for how they operate in the marketplace?

_____

_____

_____

_____

_____

_____

We can also express who we are or where we are in our life as a metaphor by using construction materials and drawing such as play-doh, clay, paint, LEGO, sticklebricks or other similar things. Sometimes when we take something we used last as children, it enables us to reconnect with the child within and discover our deeper needs that are not logical or easy to express in the adult world.

Another way of doing this is to take stones, shells, little figures such as moshi monsters or other objects you can find out on a walk. Choose objects that in some way remind yourself of you and how you feel right now. In a therapy situation, ask your client to choose things and bring them in to their next consultation. Because we aren't a stone or a feather or a shell, the chooser has to project their feelings onto it and these feelings come from the subconscious.

Exercise 69

Take a moment to explore your own metaphors.

Work is_____

Home is a_____

My relationship is a_____

My friends are_____

My family is a_____

My body is a_____

Driving in my car is_____

If I were an animal at work, I'd be a_____

When I play sport, I'm a_____

When I am happy, I am a_____

Look back at your metaphors and be curious about what they tell you about your unconscious mind. If there are any that are negative and upsetting, say for work, imagine what metaphor you'd like to have instead and how you could make changes so that the new metaphor could replace the old one.

Now let's incorporate what we know of metaphors with our learning about clean language. We will use the Lazy Jedi questions. Copy out your answers from Exercise 69 and write them in the space provided then use the clean questions for the next exercise.

_____

Exercise 70

Work is_____

- What kind of X (is that X)?
- Is there anything else about X?

Home is a_____

- **What kind of X (is that X)?**
- **Is there anything else about X?**

My relationship is a_____

- **What kind of X (is that X)?**
- **Is there anything else about X?**

My friends are_____

- **What kind of X (is that X)?**
- **Is there anything else about X?**

My family is a_____

- **What kind of X (is that X)?**
- **Is there anything else about X?**

My body is a_____

- **What kind of X (is that X)?**
- **Is there anything else about X?**

Driving in my car is_____

- **What kind of X (is that X)?**
- **Is there anything else about X?**

If I were an animal at work, I'd be a_____

- **What kind of X (is that X)?**
- **Is there anything else about X?**

When I play sport, I'm a_____

- **What kind of X (is that X)?**
- **Is there anything else about X?**

When I am happy, I am a_____

- **What kind of X (is that X)?**
- **Is there anything else about X?**

How much more do you understand about yourself from this exercise by having added in the clean questions?

Listen out for other people's metaphors and use clean questions and hypnotic language to explore how you can influence them with integrity.

Exercise 71

Now have a go at writing out your own hypnotic script. Think about what you'd like to say to yourself or to someone else. Remember to:

→ use examples of all the hypnotic language introduced in this chapter

→ write it in your own internal reference – visual, auditory or kinaesthetic

→ bear in mind your meta-programmes – choose your own chunk size, choices or process, towards or away from, match or mismatch

→ write it in the present tense

→ associate into it as if you want it now

→ use clean questions and metaphors

→ adopt an air of curiosity ('I'm wondering…')

_____

_____

_____

_____

_____

_____

_____

_____

_____

_____

_____

_____

_____

When it's done, find some background music that sounds hypnotic and record yourself reading out your script slowly, in a low, monotonous voice. Then use the recording whenever you need to. Write more scripts for other situations or issues.

Focus points

- We can use the Milton Model of deletions, distortions and generalizations to positive effect by being vague and encouraging reflection and a receptive state.
- Toxic words like 'try', 'if', 'but' and 'don't' can embed negatives into communication and interfere with influencing others to do what we want.
- Curiosity – which is key in so many NLP techniques – is even more so when influencing others with integrity because, instead of directly questioning or instructing, we are prompting them to find the response from their unconscious mind.
- Clean questions, 'yes tags' and metaphors are all devices to use in rapport when you want to encourage others to consider alternative strategies and develop enhanced communication.
- Using hypnotic language is a powerful NLP technique for influencing others with integrity.

 # Where to next?

In the next chapter, we will touch again on giving and receiving feedback. This area is crucial in so many ways. Not only do we occasionally have to give and receive feedback on a formal basis, perhaps at work or in education, but we also need feedback as a key component of self-development as we learn what works and what could be better in our life. Giving feedback with elegance and ecologically, so that no offence or hurt is taken, enables us to strengthen our personal relationships. Feedback is an essential part of sport, where we can perfect our game by learning what we need to do to improve our performance. Remember: there is no failure – only feedback.

12 Giving and receiving feedback resourcefully

A key NLP principle is finding the positive intention behind the communication, and this is never more appropriate than in the area of feedback. Whether we are giving feedback in a formal work context or informally in a social frame, how we give it and how we receive it will determine how effective it is and whether we stay in rapport. We can see the feedback we receive as criticism and reject the positive intention of its learning for us, or we can welcome it in as an opportunity to learn and develop our resourcefulness, apply the learning in our life and get different results.

In this chapter, you will be working with situations presented to you where you complete the dialogue by writing or saying what feedback you would give. You will also be given examples of feedback and have to change the wording so it improves its efficacy.

Feedback is constant: you are giving it when you're talking to someone, sending an email, texting, making a phone call, voicemail messaging and in more formal appraisal situations; and you also receive it through the same means, at work and in relationships. How you use the feedback you receive is your choice. After all, if we always do what we've always done we will always get what we've always got. If we want a different response you can accept the feedback and change to get the result you want.

How do we recognize feedback? It can come in many forms: as body language, actions or in words. How people respond to you is feedback. How you respond to them is your feedback to them. Are you giving people feedback they can learn from? In this chapter, you will learn how to give and receive feedback.

The three elements of feedback are:

- **what you are happy with**
- **what you want less of**
- **what you want more of**

→ Feedback and rapport

Giving and receiving feedback resourcefully is about being open, accepting (unconditionally and non-judgementally curious and flexible. It is therefore best given and received in rapport, so let's remind ourselves of the rules for rapport building, which are:

- **matching physiology**
- **matching the internal VAK**
- **matching the meta-programmes – chunk size, away from/towards, match/mismatch, choices/process and internal/external**

In addition, to give and receive feedback effectively we need to incorporate the following attitudes and skills:

- **curiosity**
- **a focus on the positive intention of the communication**
- **use of hypnotic language and metaphors (Chapter 11)**
- **hooking into the detail**
- **removing our own assumptions**

Let's take these one at a time.

In giving and receiving feedback it is more respectful to match the **physiology** of the person you are talking to. Sit if they are sitting, stand if they are standing and match their body posture to be in rapport. Be aware of how different cultures adopt different physiology such as personal space needs, eye contact rules, hand gestures and so on. For example, if someone steps back from you, the feedback may be that you are invading their personal space.

If the communication is written, as in a text or email, we can match by using a similar style, even the same font style and size, how they address you and close the email, sentence length, style of writing (casual, formal, etc.), use of humour or not, language (if they email you in French then respond in French – you can use Google Translate) and physical length of the email.

You will know by now how to recognize the **VAK**, whether the speaker prefers visual, auditory or kinaesthetic processes, so match their language patterns when you give feedback and when you respond to the feedback given by others. Here's an exchange showing matching of the visual representation:

> Have you had a chance to look at the charts? I've shown them to Sales and they seem to like the layout of them.

> Yes, they look great to me. I like the way you've used images of the products alongside the figures, and maybe you could also illustrate the price point.

Similarly, match the meta-programmes you detect. Imagine being small chunk and into detail and being given feedback in a huge chunk size. How could you apply the learning? Here's an example of mismatched feedback in the towards/away from meta-programme:

> I've organized a fancy-dress party for Si's 50th. It'll be great fun and we'll get in a band and ask May to do the food. She's bound to have some good ideas. (Towards)

> OK, but make sure the costs don't go out of the window and that no one is missed off the list. We don't want to put anyone's nose out of joint. (Away from)

It is important to have an attitude of **curiosity** when giving and receiving feedback. Remember that its main purpose is to give and get learning, so be curious about the learning. Instead of making assumptions about what is said and what the speaker believes, ask. Comment on what you have noticed and check it out.

In more formal feedback situations such as appraisals at school or work, being curious may be the only way to get the detailed learning you need to develop.

- Boss: 'I've noticed you look rather bored sometimes. What's the problem there?'
- Employee: 'What am I doing or saying that gives you that impression?'
- Boss: 'Well, I've noticed that you often seem to look out of the window.'
- Employee: 'I find it helps me focus. I'm not bored at all, but thank you for mentioning it. I could probably concentrate more easily if I could turn my desk away from facing the meeting room.'

The feedback has a **positive intention**, so find it! It can sometimes be a challenge to find the positive intention when feedback seems to be negative and yet the learning is still there for you to change and develop.

Exercise 72

When giving feedback, you can make the positive intention clear at the onset. One way of doing this is to use a 'feedback sandwich':

→ First layer – overall positive comment

→ Feedback – what we'd like more or less of or how it needs to be different

→ Top layer – overall positive comment

For example:

- 'I really like what you've done with my hair.'
- 'I notice one side looks a bit longer, so I'd like a bit more off on the left, please.'
- 'Thanks, that's going to look great.'

Have a go at writing some feedback sandwiches based on your own situations at home or at work.

_____

_____

_____

_____

_____

_____

_____

_____

_____

_____

_____

_____

_____

_____

We can also use hypnotic language in our feedback, which – coupled with an air of curiosity – will elicit the learning from the feedback. For example:

- **'I'm wondering why you passed the ball to Paul when I was calling for it.'**
- **'Did you mean to ignore me the other day?'**
- **'I was curious about your comment in the meeting just now; how was that relevant?'**

You can also add an element of humour to your curiosity, like this:

- **'I'm really pleased you have put your coat on by yourself and I'm wondering whether you are going to go to school with only one shoe on!'**
- **'It's great that Ben is so enthusiastic about our new concept and I'm wondering when he'll stop emailing us with questions so we can finish it for you!'**

Metaphors can also be useful in feedback, for expressing how you feel:

- **'When you kept interrupting me in that meeting I felt I was going to explode. I'm wondering whether you did it intentionally.'**

Get the **detailed** learning so that you can apply it. It's not very useful to just be told 'That's better' or 'Well done', without being given the information about how it was better. You need to know in what way better, specifically, or how it was well done, in what way.

Example

Here's an example of the level of detail that would be helpful learning for someone:

'I really enjoyed your dinner party on Saturday. The way you decorated the table was superb, with those pretty green candles and matching napkins. I appreciated you remembering that I don't like beetroot! The chicken dish you served as the main course was delicious, just the right amount of spiciness, I thought, and the dessert was amazing; those strawberries were so juicy and sweet.'

We tend to make **assumptions** based on our own values and beliefs, and these can interfere with feedback because we place our own interpretation on the words. Rather than making an assumption about someone's feedback, be curious about their beliefs and values by being an 'empty vessel'. Ask them what they mean because their perception is true for them. The meaning of a communication is what is received. We need to understand the other person's map of the world and step into their shoes to get the learning from the feedback.

Another (of the many!) presuppositions of NLP is that 'if you spot it you've got it'. This means that the attributes we recognize in others are those where we ourselves have an imbalance. Maybe we, too, demonstrate that attribute that we've found annoying or impressive in others, or perhaps we need that skill ourselves. So when you experience something in someone else that you don't like, consider how you do the same thing yourself or how you would benefit from doing it. What's happening, in a sense, is that the other person is mirroring our imbalance back to us.

Example

You notice that someone in your social group seems to avoid buying rounds of drinks by picking a moment when everyone's glass is full. You think they are rather mean. So what feedback do you get from this reaction? In what way can you sometimes be mean or are you not mean enough?

You realize that you tend to spend much more on your friend's birthday presents than they spend on you. Maybe the learning here is to correct this imbalance.

The following exercises will give you a chance to have a go at putting these ideas into practice.

Exercise 73

Think about a recent time when you have been annoyed by someone's actions or what they said. Write down in the space below what happened.

_____

_____

_____

_____

Now think about what learning you can take from the feedback. What have you learned about yourself?

_____

_____

_____

_____

What can you do differently as a result?

_____

_____

_____

_____

What we are learning here is that before you think about giving negative feedback to someone else, just run it past yourself first and ask yourself, in what way do I have the same structure? What can I learn from my reaction to them and how can I use this reaction to learn a different way of being?

Exercise 74

Think about a recent time when you have been impressed or touched by someone's actions or what they said. Write down in the space below what happened.

_____

_____

_____

_____

Now think about what learning you can take from the feedback. What have you learned about yourself?

_____

_____

_____

_____

What can you do differently as a result?

_____

_____

_____

_____

When you notice something you admire in someone else this means either that you have it, too, in some way, or that you have an imbalance in that area. What you recognize in others you are capable of yourself because, through noticing it, it means you have the structure for the skill and can release the potential.

_____

Exercise 75

Here are some examples of poor feedback. Rewrite them so that they are resourceful as a learning experience.

→ What time do you call this? You're late again.

_____

_____

→ I wish you'd stop interrupting me.

_____

_____

→ Could you slow down, we'll have an accident.

_____

_____

→ You're boring.

_____

_____

→ You smell.

_____

_____

→ Do you have to make such a fuss about everything?

_____

_____

→ You're in my parking space.

_____

_____

→ You're so rude when you're drunk.

_____

_____

→ You're talking rubbish as usual.

_____

_____

→ That's better.

_____

_____

Exercise 76

Giving feedback

Think of some feedback you would like to give someone and write it down here.

_____

_____

_____

_____

_____

_____

First, let's remind ourselves of the top tips for giving feedback.

→ Check you are in rapport.

→ Ask yourself how this feedback is also true for you.

→ Create a well-formed outcome of both of you learning from the feedback.

→ Tell them what you have noticed.

→ Maintain eye contact and imagine them accepting the feedback.

Who do you want to give feedback to?

_____

_____

What do you want to say?

_____

_____

Now write out a script using the tips listed above.

_____

_____

_____

_____

_____

_____

Receiving feedback

For completeness, here are the tips for receiving feedback.

→ Be curious and open to feedback constantly.

→ Remind yourself that this is learning and welcome it in.

→ Use the present tense.

→ Presuppose they are right.

→ Respond by saying, 'In what way do I...?'

→ Stay in rapport.

Now imagine that you are receiving that feedback. How will you respond?

_____

_____

_____

Focus points

- There is no failure, only feedback. The positive intention of feedback is that it is learning, so reframe feedback in that context.
- We cannot 'not communicate': before we have even said anything, we are communicating something, because 93 per cent of communication is non-verbal. The meaning of what is communicated is the message so, regardless of what we meant to communicate, how it is taken is what it is.
- We can learn from feedback – what works and what could be improved – and we can pass that learning on as well, by giving feedback to others in such a way that they can use the feedback as learning.

- We give and receive feedback constantly at work, at home and in our social life, in our sport and our relationships. It is important not to make assumptions but to be an empty vessel and be curious about what we observe rather than feed back our assumptions about what we think someone meant.

- We need to use our 'chunking skills' to chunk down to the detail to get the essence of the feedback. Since we also have the structure of whatever we noticed in someone else, we need to be curious about that and ask, 'How is that true for me?'

 # Where to next?

In the next chapter, we will explore how the mind and body are one, how what you believe and think is what you get. You will have touched on this in Chapter 10 on modelling when you learned that you couldn't get the skill without the belief. Although this concept is very useful in the area of exercise and sport, it is also very relevant for body image (slimming) and health. Maybe you have noticed that people who are not happy at work take more sick days than those who are fulfilled and enjoy their job? Healthy people tend to have 'just a cold' and unhealthy people are 'suffering from flu'. What you believe has a significant effect on what you get.

13 The mind–body connection

Sportspeople often say that the difference between success and failure in sport is the difference between your ears! In other words, it all happens in your brain and your thinking. This is true of all areas of our life. In NLP, mind and body are considered to be one system, with each directly influencing the other. You can change the way you feel by what you think and, in the same way, what happens in your body affects your thoughts. When you change one part of the system you will have changed another. This chapter explores how we can make our body work better through some physical exercises. We will also learn how we can use this connection for a healthy life and to achieve the body we want.

Neuro-linguistic programming is a study of the structure of excellence. How we think affects what we say and what we do. If we want to improve what we do, we have to change the way we think. If we always do what we've always done we will always get what we've always got. If you want a different result in your sport, you need to change the way you think and what you do. Your mind and body are one. Each thought you have transmits to the furthest cell in your body through neurotransmitters and the same neurotransmitters that you have in your brain can also be produced in your internal organs. The mind and body work together as an integrated whole.

 ## Exercise 77

Check in with how you are feeling right now. On a scale of 1 to 10, how confident do you feel? ☐

Now give yourself a little shake, stand up, stand tall, head up and look straight ahead. How confident do you feel now? ☐

What difference did you notice when you changed your posture?

_____

_____

_____

_____

_____

→ The inner game

 '[The inner game] is the game that takes
place in the mind of the player, and it is
played against such obstacles as lapses in
concentration, nervousness, self-doubt and
self-condemnation. In short, it is played to
overcome all habits of mind which inhibit
excellence in performance.'

W. Timothy Gallwey, *The Inner Game of Tennis*

Timothy Gallwey, who used the word 'inner' originally in 'the inner game of tennis', 'inner skiing' and 'the inner game of golf', urges us to become consciously unconscious and take the busy thought processes out of the equation so that the mind becomes one with the body and the body functions work automatically without interference from thoughts. He especially encourages us to be non-judgemental: 'When the mind is free of any thought or judgement it is still and acts like a perfect mirror.'

Exercise 78

This exercise aims to show you how to put into practice Gallwey's ideas.

* First, check your body and give yourself a score for how tense each muscle is feeling. Put a score out of 10 for your shoulders, back, hips, arms and legs where 1 is not tense at all and 10 is very tense.

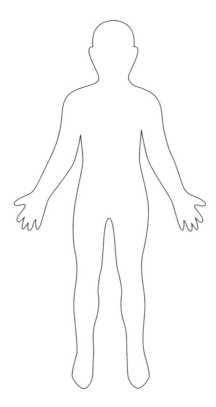

Now just close your eyes and stop thinking. Imagine your thoughts are butterflies or birds, something that can fly off. Just send the thoughts out of your mind. Do this several times and build up the length of time you can keep your mind calm.

Concentrate on your breathing. Think about it as you breathe in and out. Notice the rhythm. It may help to allow your hands to open as you breathe out and close as you breathe in. Notice if your stomach is expanding as you breathe in and contracting as you gently push the air out. Put your hand on your stomach and check. You will breathe more effectively if you pull your stomach in to breathe out and let it expand as you draw the air in.

Now rescore the tension in your muscles.

What did you notice during this exercise?

_____

_____

_____

_____

How could you use this to manage your state at work, at home or during your sport?

_____

_____

_____

_____

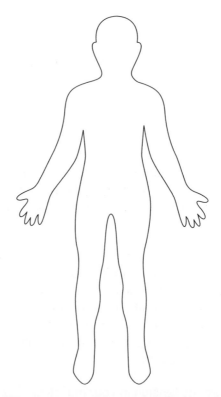

In some sports you can use your breathing to good effect by breathing in before you take a hit and breathing out as you strike the ball. Think about how you could use your breathing like that in your sport.

_____

_____

To get the best performance we have to believe that we have the potential for it, that we already have that resource for competence in the sport. This is fundamental to NLP – we already have all the resources we need to do whatever we choose to do. We have the resources to learn how to do something and if someone can do it, we can learn how to do it, through coaching and modelling excellence. We also have to choose to do it and choose to believe we can do it. Believe the best is in there.

Your unconscious mind allows your body to breathe, move and respond to temperature, and in sport you will run to hit the ball or kick it instinctively without conscious thought. However, in order to get a different, better result the unconscious needs to become conscious so that we can analyse and then tweak those unresourceful limiting beliefs.

Let's start off with some experiments that will show you just how powerful your mind is in controlling your body.

Exercise 79

It may be easier if you ask your partner or a friend to read out the instructions while you do these exercises.

Stand and put your right arm straight out in front of you with your finger pointing.

Remain facing forward and move your arm as far around to the side and, if you can, behind you, keeping your arm straight. Notice the point you reach with your finger. What are you pointing at?

_____

Now close your eyes and repeat the exercise and this time visualize (imagine) going farther around, staying relaxed and comfortable.

Now imagine getting a little farther, again stay relaxed and comfortable.

And then go another few inches, again staying relaxed and comfortable.

Now move your finger another few inches.

Open your eyes and see how much further you have managed to reach when you use the power of your mind.

_____

Exercise 80

Here's another exercise to show how you can influence your body by using your mind. Athletes and sportsmen do this all the time.

Put out your right arm, palm down.

Your partner rests his hand lightly on your arm. He doesn't need to push on it, just enough pressure to give you something to push against.

Test it by pushing your arm up and notice the effort it takes.

Now think of something that hasn't gone well for you recently. Think about it as if it's happening now. Then raise your arm and notice how much harder it is to push your partner's hand up.

Break state for a moment by shaking your arm and then return to holding it out again with your partner's hand on it as before.

Now think of something that's gone really well recently and think of it as if it's happening now. Then raise your arm again and notice how much easier it is to push your partner's hand up.

Exercise 81

Make a circle with your left finger and thumb.

Now link your right finger and thumb through the circle you have made to form a link like a chain so you can only unlink by pulling on one or other of your hands.

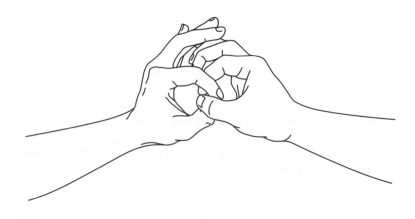

Think of someone you really like and then pull hard to try and break the chain.

Now think of someone really dislike and pull hard to break the chain.

Wasn't it much easier to break the chain when you were thinking of someone you disliked?

See how belief can affect how your body responds. If we believe we have a poor backhand or that we are not good at running, putting or some other physical task, we tend to notice when this is so and we delete, generalize or distort examples that don't fit this belief. We say, 'Our backhand shots are always useless', or 'This means my backhand is weak'. When we hold this belief in our head, the body responds accordingly, when we can – by calming our thoughts and being free of judgement (good and bad) – just trust our body to work.

- Visually referenced players can improve the way their body works by focusing on what they see, where they are aiming for, the ball, the place they are running to, visualizing the finish line and so on.
- Auditorily referenced players may work on listening to the sound of the ball on the racquet or stick, the sound of the pedals or of trainers on the ground.
- Kinaesthetically referenced people will focus on the feel of their body, how the muscles feel and the rhythm of the play.

If you want a slimmer, more toned body the same focus applies. Visualize how you want it to look if you are visually referenced. Imagine what your friends and family will say when you have achieved it if you are auditory and how your clothes will feel loose on you if you are kinaesthetic.

Awareness needs to be totally present in the here and now with every shot, every point. By staying in the present, we stay calm. Anxiety and stress are almost always about what 'might' happen or what has happened being generalized into a belief about what will happen. 'I've played that shot badly so that means I'll lose this game.'

Notice how your physiology affects your performance. Acting 'as if' you are the best you can be in terms of adopting the physiology will get the desired state. Notice how your models of excellence hold themselves.

Example

Judy wanted to be as slim as some of the people she worked with, so she modelled how they chose food in the work canteen and how they ate. She noticed that simply choosing the same food was not getting her the model. She even ate as slowly as they did and chewed each mouthful carefully. She drank water before eating rather than with the meal or afterwards and she stopped eating when she felt she'd had enough. All these things helped her and she noticed that her clothes were beginning to feel looser. However, the difference that made the difference was when she copied the way her slim colleagues walked and sat. They moved more actively and acted as if they knew they looked good, whereas Judy tended to slouch and walk slowly. As soon as she matched the physiology she started thinking like a slim person and made slim person's choices quite unconsciously.

We can also change our state by using our eyes.

- Look up and to your left (visual remembered: Vr) to remember images from the past, when you played well, did something with excellence and bring it to the present.
- Look up and to your right (visual constructed: Vc) to create the image you want, the image of you winning the game, playing a masterful shot, scoring a goal.
- Straight ahead but body weight to the left (visual remembered).
- Straight ahead and body weight to the right (visual constructed).
- Look to your left (auditory remembered: Ar) to recall what someone has said.
- Look to your right (auditory constructed: Ac) to imagine what someone might say.
- Look down and to your left (auditory internal or auditory digital: Ai) for inner dialogue.
- Look down and to your right (kinaesthetic: K) to access feelings.

Exercise 82

Think of something you'd like to change and write it down in the box below.

Now use your eyes to see what happens to this when you move them to a different position. Here's an example of how this works for weight loss.

→ Look up and left to recall how I used to look before I put on weight.

→ Look up and right to visualize how I'd look if I were a few pounds lighter.

→ Look left to recall what people said to me when I lost weight last time.

→ Look right to imagine what they will say when I am slimmer.

→ Look down and left to tell myself I can do it again.

→ Look down and right to feel motivated to do something about my weight.

Now you do it with the issue you'd like to address and notice which eye movement makes the difference to you.

Look up and left to recall

Look up and right to visualize

Look left to recall what was said to me when

Look right to imagine what they will say

_____

_____

Look down and left to tell myself

_____

_____

Look down and right to feel

_____

_____

We have learned that we can use our body to change our state through our breathing, eye cues and physiology. Because our mind and body are one, we can also influence what happens with our body by thinking differently. We have learned already in previous chapters how we can examine and change limiting beliefs, so now is your opportunity to challenge limiting beliefs about your body in connection with how you feel about it and what you can achieve: a better shape, more fitness and energy, better movement and sport and exercise skills.

Exercise 83

What belief would you like to change about your body? This could be in relation to illness, sport, appearance or exercise. Write down your limiting belief in the box below.

```
┌──────────────────────────────────────────────┐
│                                                │
│                                                │
│                                                │
│                                                │
└──────────────────────────────────────────────┘
```

Now think of a time when you did not have this belief. Use the timeline (see Chapter 6) to take yourself back to a time when you had a more resourceful belief about yourself. Stand on that place on the timeline and anchor it.

To anchor this more empowering feeling and experience, decide on an action such as squeezing your earlobe as you associate into the memory as if you are feeling it right now. The more you imagine you are feeling this much more resourceful feeling relating to your body, keep the anchor in place (the action you're associating with it).

Do you notice a change in your physiology as you associate into this feeling?

You can use this anchor whenever you want to change your belief about this issue.

Focus points

- It has been well documented that exercise increases our endorphins or 'happy hormones', making us feel more upbeat and positive, enabling us to concentrate and work better as a result and generally keeping us feeling happy and healthy.
- Your mind and body are one. Notice and increase your awareness of this by changing your belief, adopting a resourceful physiology and keeping your mind free of judgement.
- As you live in the moment, you can achieve what you desire in all matters concerning your body.
- Understanding the mind–body connection means that you can use your body to change your state of mind; since mind and body are one and the same, there are ways to change your thinking so that you get a different result in your body.
- This will be apparent if you do sports or another activity and use the modelling techniques from Chapter 10. You will notice the difference between simply copying what someone does and getting the belief that goes with it, then copying the action.

 ## Where to next?

The final chapter covers the TOTE (Test Operate Test Exit) technique, which is how we check that our strategies are working for us. We will also look at state management, which is how we calibrate, change and recalibrate how we are, our state of mind. Being able to manage our state so that we can be resourceful and effective whenever we need to be is another useful learning tool described in the next chapter.

14 Strategies for success

In this final chapter we will learn how to use the TOTE (Test Operate Test Exit) to apply the NLP skills and techniques we've learned in the book and check that they are working. The TOTE is the way we constantly test our strategies for success. We will also learn how to use the timeline to revisit and change perceptions of memories and past events that hold us back from achieving our full potential.

→ The TOTE exercise

The TOTE works on the principle that we have a compelling outcome. This is our goal, what we want to achieve from a specific behaviour. It is based on the principle that we learn more from our mistakes than from successes and when we feed back the learning from failures we improve and gradually get closer to what we want to achieve. The learning moves us from conscious incompetence through to unconscious competence. At every stage, we are comparing what we have with what we want. But how will we recognize it when we have it? What evidence do we need? The TOTE is essentially a feedback loop; without evidence, we could just keep going around and around the loop!

TEST:
Has the goal been achieved?
← If YES – EXIT
If NO – OPERATE (DO SOMETHING DIFFERENT) AND TEST AGAIN →

1 Test (1)

This is the comparison between what we have and what we want. It is the difference between what we call in NLP 'the present state' and what we want, which is 'the desired state'.

2 Operate

Here we apply our resources, remembering that we already have all those we need.

3 Test (2)

Compare again what we have and what we want. If it is a match, we can exit. If not, go back and apply different resources. Repeat until the desired state is achieved.

4 Exit

Once the desired state or compelling outcome is achieved, we can exit.

→ Strategies

A strategy is how we do what we do to get the desired outcome. Strategies are habits initiated by a trigger that prompts a series of steps and ends with an outcome. We can use the VAK internal and external symbols to help us here. Be aware, as you think it through, of the following:

- **what you visualize (Vi)**
- **what you see (Ve)**
- **what you hear internally (your inner voice) (Ai)**
- **what you hear externally (Ae)**
- **what you feel (Ki)**
- **what actions you take (Ke)**

For some strategies it may also be relevant to record what you smell or taste (O).

When you are eliciting your strategy, think about what you do as if you are doing it now, in the present time.

Example

One example of evaluating a strategy is 'how I get dressed in the morning':

- I visualize my day: Vi (visual internal).
- I think about what I will be doing: Ki (kinaesthetic internal).
- I look at the clothes left on my floor: Ve (visual external).
- I pick up something appropriate: Ke (kinaesthetic external).

Now this doesn't always work because I get dressed when it's dark outside, so sometimes I find that it's colder than I expected, or I've put on tennis gear but it is raining so I can't play tennis. Therefore a better strategy might be to:

- ask my husband what the weather's like: Ae.
- decide what I'll be doing: Ki.
- select appropriate clothes: Ke.

Exercise 84

Decide which strategy to evaluate.

_____

_____

Write it down in the same way as the example above.

_____

_____

_____

_____

_____

Is it effective? Does it meet your desired outcome?

_____

_____

If it does – exit.

If it doesn't – consider other, more effective strategies (operate and test again).

Write them down and try them out.

Do they work?

If so, exit.

If not, have another go.

We run TOTEs unconsciously all the time, as we check our present state against our desired state and make small adjustments until we have a match. This feedback loop prompts you to find the strategy that works to get what you want. Criteria for success are:

- **having a well-formed outcome that is clearly defined and achievable**
- **belief that your outcome is of value to you**
- **belief that you have the potential to achieve it**
- **awareness of what is happening at each step of the strategy**
- **willingness to learn from the feedback**
- **willingness to be flexible and try everything**
- **willingness to model other people's strategies**
- **willingness to keep going until you reach your goal**

If you have completed all the exercises in this workbook you already have all these skills. The TOTE is a way of consolidating this learning as part of your ongoing personal development.

In the chapter on modelling you learned how to acquire models of excellence from others. You can apply the TOTE as you test the model to find out which steps in the strategy make the difference in getting the desired outcome, the model you need to access the skill.

Exercise 85

One of the things that can hold us back from having a well-formed outcome is a lack of belief in our potential to achieve it. We need to see that perhaps we have achieved something similar in the past and for this the timeline can be very useful.

Imagine your timeline and stand on the place on the line that represents the point in your life when you were able to do or be what you want for yourself now.

Associate into that feeling and see, hear and feel what you felt then about this thing, as if it were happening today. When you are totally associated, anchor that feeling of possibility.

Still using this anchor, walk to the point on the timeline that represents today for you and imagine you are now doing and being the person you are, with the belief anchored. How does that feel?

Sometimes we have a memory that holds us back from being able to do something, or a negative experience that has stayed with us and is limiting our choices today. If that is the case for you, go back to the timeline and stand on the point when this happened.

You now need to disassociate by flying above the timeline in your imagination and looking down on yourself at that point. What do you notice about the person standing there at that point in your life? What do you see, hear and feel about them then?

Now imagine you are floating above the point that represents today and look back at that person. What could you tell them about what you observe? Does the belief they hold work for them today? How could they change it to be more resourceful?

The timeline is there for you to use whenever you need to and the more you use it the more proficient you will become at moving up and down through time and hovering above to observe yourself from what we call the 'third position', which is disassociated like an impartial observer. When you get really good at using the timeline, you won't even need to imagine the line on the floor but you will be able to use it in your imagination.

Exercise 86

The point on the timeline that represents today is sometimes called your state. It can be a mood or feeling about how things are for you right now. It is useful to calibrate this so you can recognize it for yourself.

Start by calibrating your neutral state. How are you now? Be aware of your physiology, your facial expression, your body temperature and the sound of your voice.

Now let's calibrate a more upbeat state and then anchor it so you have access to it when you need it. This is much healthier than chocolate or a glass of wine! When you are happy, feeling confident and resourceful, how do you look? What do you sound like? What do you feel and what do you do? Associate into that feeling and be aware of your physiology: what does your body look like? You can check in a mirror, which may help. Say a few words and notice how your voice sounds. Where is the good feeling inside you? Can you touch the place where you feel good? Anchor it so that you have an action to associate with that state.

Now think of something that makes you feel sad. Go through the same process and be aware of how your physiology changes. Remember that mind and body are one, so the sad thoughts will be reflected in your body. There's no need to anchor this one!

You may want to do the same exercise for angry, if this is something you want to manage, so that you can recognize when you are getting into that state as well.

Managing your state is about being aware of your mood at all times so if it is not resourceful you can change it by altering your physiology. By anchoring the desired state, you can access it when you need it. It can be very useful to calibrate states in other people; indeed, you may already do this instinctively. Can you tell what state your boss is in by any outward signs, their physiology, and any habits they have that indicate how they are feeling? You can probably read your partner's moods and your child's but are you aware how you calibrate them? Make the unconscious conscious so you can use this skill on people you don't know well.

Focus points

- This book contains all the NLP tools and techniques you need to enable you to be the resourceful, excellent person you are. You will need to practise them so that they become automatic and natural. Use them in all areas of your life and be a model of NLP for those around you, your peers and your children.
- Remember that you already have all the resources you need, so notice when you use them and think about how else you can use a particular resource. Observing others is a good way to learn more about yourself because you will notice qualities in others that you have yourself.
- Feedback is a valuable source of learning, so be brave and ask for feedback from your colleagues and friends. Be prepared to give feedback to others and this will encourage them to return the favour because this is what it is.
- Feedback is learning, so bring it on! In fact, please leave feedback for the author on Amazon or whatever book site you use or any business networking groups and social networking groups.
- There are numerous other books on NLP, many of them specific to business, sport, children, relationships and so on. The tools and techniques in this book adapt to all areas of life, but if one particular aspect of NLP interests you, then explore it in depth by further reading.

 # Where to next?

NLP is experiential and you will learn more about how it works by going on a practical NLP course where you will have the chance to work with others and with NLP trainers who are experienced at observing things that we sometimes don't notice ourselves. If you have been inspired to become an NLP practitioner, contact the author who can either offer a course to suit you or recommend one in your area.

Bibliography

Bavister, S. & Vickers, A., *Teach Yourself NLP* (London, Hodder Education, 2008)

Gallwey, T.W., *The Inner Game of Tennis* (London, Pan Books, 1975)

Knight, S., *NLP at Work* (London, Nicholas Brealey Publishing, 2009)

Lazarus, J., *Ahead of the Game* (St Albans, Ecademy Press, 2006)

O'Connor, J. & Seymour, J., *Introducing NLP* (London, Harper Element, 2002)

Ready, R. & Burton, K., *Neuro-linguistic Programming for Dummies* (Hoboken, Wiley, 2004)

Ready, R. & Burton, K., *Neuro-linguistic Programming Workbook for Dummies* (Hoboken, Wiley, 2008)

Whitmore, J., *Coaching for Performance* (London, Nicholas Brealey Publishing, 2009)

Website: SLD International, www.leaderperfect.com, has a useful article on NLP and the brain by Dr Mike Amour. Go to the training section of the site and click on NLP.

Index